Multi-Gigabit Microwave and Millimeter-Wave Wireless Communications

For a list of recent titles in the
Artech House Mobile Communications Series,
turn to the back of this book.

Multi-Gigabit Microwave and Millimeter-Wave Wireless Communications

Jonathan Wells

ARTECH
HOUSE

BOSTON | LONDON
artechhouse.com

Library of Congress Cataloging-in-Publication Data
A catalog record for this book is available from the U.S. Library of Congress.

British Library Cataloguing in Publication Data
A catalogue record for this book is available from the British Library.

Cover design by Merle Uuesoo

ISBN 13: 978-1-60807-082-4

© 2010 ARTECH HOUSE.
685 Canton Street
Norwood, MA 02062

10 9 8 7 6 5 4 3 2 1

Contents

Preface

Since my earliest engineering days, I have had a keen interest in both high data rate and high-frequency communication systems. Back in 1985, when I was an undergraduate student working at the Plessey Research Labs in Caswell, U.K., I developed a 167 Mbps fiber-optic prototype—a very fast system at the time. In 1990 I was awarded a doctorate for novel receivers operating at 94 GHz, and afterwards undertook postdoctoral work building 183-GHz receivers, modeling 600-GHz devices, and even considered a grant proposal developing 1.2-THz devices. I remember how disappointed I was when I moved into industry and realized commercial applications were several orders of magnitude slower and lower in frequency.

Things are very different now. Microwave radios in the 6- to 40-GHz bands now form a significant part of many countries' telecommunications infrastructure. Over 1 million such terminals were shipped in 2009. The fastest-growing segment of this market is high data rate radios, supplying high-speed connectivity for cellular backhaul, emerging "next generation networks," and business-to-business enterprise connectivity. Recent advances in signal processing and RF technologies have enabled wireless links to be engineered to reach gigabit per second speeds. Newer technologies at the 60-GHz and higher millimeter-wave bands take advantage of new wireless regulations specifically designed to enable ultrahigh capacity communications, allowing previously unheard of data rates of 1 Gbps and above to be economically transmitted between wireless cell sites for telecom applications, or around the house for HDTV distribution or personal home network applications.

I now work as a consultant in this new and exciting field, advising clients on various technical, marketing, regulatory, and legal aspects of high data rate, high-frequency communications. Since there are currently no technical texts available that cover this subject area, I wrote this text to provide a thorough review of the latest multi-gigabit per second wireless technologies, their capabilities, applications, and limitations for ultrahigh data rate wireless transmission. Although this book includes a number of academic and theoretical sections, I have also tried to place emphasis on practical use and applications, particularly on regulatory approval, system design, and link planning to determine operational performance given the physical limitations of operating with high data rates at high frequencies. I hope this book balances theoretical content with practical, real-world advice based on real-world experience.

Acknowledgments

I am grateful to the many people, who have helped develop and foster my interest in high-frequency, high data rate wireless communications, and assisted my career progression within this field of study. Of particular significance I would like to acknowledge Prof. Nigel Cronin at the University of Bath for first nurturing my interest in millimeter-waves, and both Paul Kennard and Doug Lockie and their colleagues at Stratex Networks (now Aviat Networks) and GigaBeam, respectively, for showing me how such technologies are developed, built, and commercialized in the real world. I would also like to acknowledge the many people, too numerous to mention, who have provided me useful technical insight while writing this book, and the companies and organizations that contributed photographs and illustrations. I would also like to thank Trevor Manning, author of the classic *Microwave Radio Transmission Design Guide* (Artech House, 1999) for much helpful advice when contemplating writing this book. Finally I would like to thank my family: my parents for their lifelong interest in my work, and in particular my wife, Andrea, for her confidence, encouragement, support, and continuous belief in everything I do.

1

The High Data Rate Wireless Environment

1.1 Introduction

Over the last few decades, communication technologies have advanced significantly. Transmission speeds unheard of just a few years ago are now commonplace. Most computers now ship with a 1 Gbps Gigabit Ethernet (GbE) port as the standard networking interface. GbE has become the de facto protocol for data transmission in consumer applications. In the industrial space, 10 Gbps Ethernet (10 GbE) is widely used. 10 GbE transmission forms the fiber-optic core of telecom networks, and is now a standard backplane for ultrafast commercial equipment enclosures and specialized server blades, switches, and routers. Even the 10-Gbps solutions available today are insufficient for future network needs. 40-Gbps solutions are becoming available, which multiplex four of these 10-GbE carriers together. Single carrier 40-Gbps and 100-Gbps (40 GbE and 100 GbE, respectively) Ethernet standards are under development by the IEEE 802.3 Higher Speed Study Group (HSSG) under the IEEE 802.3ba designation.

Such enormous data rates are readily supported by fiber-optic cables, which deliver such speeds across massive global networks. Although fiber is an excellent transmission medium, its availability is limited and such high-speed services are expensive to roll out and lease. Studies have shown that in 2009, only 23% of businesses in the United States with 20 or more employees have access to broadband services over fiber. The number is worse in Europe with only 15% of businesses connected via fiber. Most large enterprises are connected, but small to medium businesses are vastly underserved. Even when connected, high data rate services are expensive to lease. In the United States, the average lease rate for a GbE broadband connection is $4,300 per month with a multiyear commitment. Rates are generally higher than this in Europe. Installing dedicated fiber services is usually not a viable option, as trenching costs can reach $250,000 per mile in U.S. metropolitan areas (assuming fiber trenching is allowed), and planning and impact studies can add many months or even years to rollout schedules. In many places, fiber simply cannot be laid without significant disruption or environmental impact.

For these reasons, there has long been interest in wireless communications as a high-speed alternative to wired solutions. However, for a long time, wireless transmission speeds have lagged a long way behind their wired counterparts, making it difficult to offer wireless as a credible high-speed alternative to fiber. In the last few years, however, wireless speeds have begun accelerating and have significantly reduced the gap with wireline capabilities. Since the mid-2000s, wireless equipment capable of 1 Gbps and faster speeds have been commercially available using a variety of different technologies. In the popular microwave bands, equipment has been developed that employs sophisticated modulation schemes and techniques to reuse frequency channels to increase throughput dramatically. New frequency allocations in the higher millimeter-wave bands have encouraged ultrahigh data rate usage of the higher regions of the spectrum. For these reasons, wireless is now becoming a credible high data rate alternative to high-speed wireline solutions.

1.2   The Electromagnetic Spectrum

The electromagnetic spectrum is the range of all possible electromagnetic radiation frequencies. The electromagnetic spectrum is theoretically infinite, although in reality it extends from the shortest wavelength (defined in quantum mechanics as the Planck length) up to the size of the universe.

The practical section of the electromagnetic spectrum is shown in Figure 1.1, including a major part of the spectrum useful for wireless transmissions. In practice, commercial wireless systems cover the bands from a few megahertz up almost 100 GHz. A list of typical applications in these bands is also included.

The term radio frequency (RF) is a generic term used to cover all the bands, both high and low, that are used for radio transmission services. Traditionally, microwave frequencies are considered as occupying the SHF (super high frequency) band, and millimeter-waves as occupying the EHF (extremely high frequency) bands, as defined by the ITU [1]. Over the years, the UHF (ultrahigh

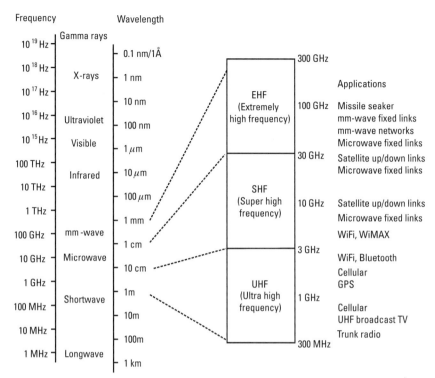

Figure 1.1 The electromagnetic spectrum (left), the radio spectrum applicable to medium and high data rate communications (middle), and typical wireless applications at these frequencies (right).

frequency), SHF, and EHF bands have been subdivided into a number of smaller bands, to enable association with more precise frequency sub-bands and applications. This was driven primarily by the growth of radar, and widely used designators were allocated by the IEEE [2]. Other band designations have been derived over the years by a variety of groups including the European Union, NATO, and the Radio Society of Great Britain. Some are widely accepted, and others used only in specialist cases. A summary of the major band designations in common use are depicted in Table 1.1.

It should be noted that strictly speaking, the term millimeter-wave refers to EHF transmissions with a wavelength of less than 1 cm, or a frequency of 30 GHz and above. However, in the wireless communications world it is more convenient to refer to millimeter-waves as operating at higher frequencies. This is because the globally available and widely used bands from 6 to 40 GHz, commonly called the microwave bands, are relatively consistent in characteristics and are managed in a similar way by regulators around the world. However the bands at around 60 GHz and higher have very different atmospheric propagation characteristics and are thus managed and regulated very differently by regulators. It

Table 1.1
Common Band Nomenclatures for the Radio Frequency Spectrum

ITU Nomenclature [1]			IEEE Nomenclature [2]		Other Nomenclatures	
Band Designation	Band No.	Frequency Coverage	Band Designation	Frequency Coverage	Band Designation	Frequency Coverage
VHF	8	30–300 MHz	VHF	30–300 MHz		
UHF	9	300–3,000 MHz	UHF	300–1,000 MHz		
			L-band	1–2 GHz		
			S-band	2–4 GHz		
SHF	10	3–30 GHz				
			C-band	4–8 GHz		
			X-band	8–12 GHz		
			Ku-band	12–18 GHz		
			K-band	18–27 GHz		
			Ka-band	27–40 GHz		
EHF	11	30–300 GHz			Q-band	30–50 GHz
			V-band	40–75 GHz	U-band	40–60 GHz
					E-band	60–90 GHz
			W-band	75–110 GHz		
					F-band	90–140 GHz
			mm	110–300 GHz		
					D-band	110–170 GHz

is therefore convenient to separate these bands from the lower frequency microwave bands. Throughout this book, the term microwave bands is used to refer to the 6- to 40-GHz bands, and the term millimeter-wave is used to refer to the bands of around 60 GHz and higher.

1.2.1 Management of the Radio Frequency Spectrum

The RF spectrum is a scarce resource. In economic terms, a scarce resource is something that once it is used up or destroyed, it cannot be recreated or reused. This is true for RF spectrum, since once a user is operating at a particular frequency in a certain geographic area, no other users can generally use that same frequency without the risk of interference. For this reason, sensible allocation and coordination of the RF spectrum is a necessity, allowing users and operators to coexist with one another and offer different wireless services that share the RF spectrum harmoniously.

There are a number of groups who have been specifically set up and tasked to allocate, coordinate, and manage the RF spectrum around the world.

1.2.1.1 ITU

The International Telecommunication Union (ITU) is the primary regulatory agency in the world with the responsibility for coordinating the shared global use of the radio spectrum. The ITU was founded in 1865, as 20 European countries realized that the then rapid expansion of telegraph networks required nations to agree on common standards of international interconnections. Today the ITU includes over 190 member countries. The ITU is based Geneva, Switzerland.

The ITU is organized into three sectors dealing with telecommunication standardization, telecommunication development, and radiocommunications. The ITU Radiocommunication Sector (ITU-R) has responsibility for the global management of the radio frequency spectrum, for applications such as fixed, mobile, broadcasting, amateur, space research, emergency telecommunications, meteorology, global positioning systems, environmental monitoring, and communication services.

Approximately every 4 years, the ITU-R holds the World Radiocommunication Conference (WRC) to review, and, if necessary, revise the Radio Regulations—the international treaty governing the use of the radio frequency spectrum. The WRC agendas are set 4 to 6 years in advance, and finalized 2 years before the conference. Agenda items are usually based upon recommendations made by previous world radiocommunication conferences. Therefore it takes many years for introduction or modifications to spectrum plans and use. The recent introduction of some of the millimeter-wave bands discussed in this book have taken over a decade to be written into the Radio Regulations.

1.2.1.2 FCC

The Federal Communications Commission (FCC) is responsible for regulating U.S. radio, television, wireline, and wireless communications. The FCC is directed by five politically appointed commissioners, one of whom the U.S. president appoints as chairman. The FCC was established by the 1934 Communications Act and is based in Washington, D.C.

The FCC is organized into many bureaus and offices. Of interest to wireless communications are the Office of Engineering and Technology (OET), which allocates spectrum for nongovernment use and provides the FCC advice on technical issues, and the Wireless Telecommunications Bureau (WTB), which regulates the U.S. radio spectrum.

1.2.1.3 CEPT

The European Conference of Postal and Telecommunications Administrations (CEPT) was established in 1959 to manage all European post and telecommunications issues. CEPT's primary role is to facilitate European regulators and deploy ITU policies in accordance with European goals. CEPT does not provide formal policy; rather, it provides recommendations on telecom matters. Since CEPT includes 48 members covering almost the entire geographical area of Europe, CEPT's guidance is widely adopted as policy within Europe.

For wireless matters, CEPT established the Electronic Communications Committee (ECC). This has several subcommittees charged with managing activities such as European frequency management and spectrum engineering. Given the geographic characteristics of Europe, with many closely spaced, relatively small countries, crossborder propagation and wireless interference is a major concern. The ECC's frequency coordination and harmonization activities are therefore very important for the harmonious use of spectrum in Europe. The ECC's activities are managed by the European Communications Office (ECO), based in Copenhagen, Denmark.

1.2.1.4 ETSI

In 1988 CEPT created the European Telecommunications Standards Institute (ETSI), into which all its telecommunication standardization activities were transferred. ETSI is now responsible for preparing and issuing standards under the European Commission's Radio Equipment and Telecommunications Terminal Equipment (R&TTE) directive. These harmonized standards are mandated across Europe, and require compliance before any equipment can be sold, deployed, or operated within the European Union. ETSI is organized with several technical committees and partnership bodies who produce technical standards for fixed, mobile, radio, broadcast, and Internet technologies. These standards are recognized and mandated not only by European Union countries, but also by many radio agencies worldwide. ETSI is based in Sophia Antipolis, France.

1.2.1.5 Regional Regulators

Like the FCC in the United States, each country has its own national regulator to implement and manage national radio communications, within the framework of ITU policy and other body's guidelines or recommendations. In the United Kingdom, for example, Ofcom (the Office of Communications) is the independent regulator and competition authority, with responsibilities for television, radio, and telecommunications, in addition to wireless communications services. Ofcom ensures the U.K. wireless industry follows ITU policy, CEPT band plan recommendations, and ETSI technical rules. They also set the national licensing regulations and determine the fee structure for British radio licenses and regulated wireless services.

Similar national and/or provincial regulators exist in all regions and countries in the world.

1.2.2 Radio Frequency Plans

An organization body has several goals when drawing up frequency plans and usage rules. A key requirement is for efficient use of the limited radio spectrum, allowing users to harmoniously coexist, with minimal potential for interference. For this reason, careful consideration is placed on channel sizing and selection.

A generalized radio frequency band plan is shown in Figure 1.2. This is known as a frequency division duplex (FDD) band and is common for higher frequency and higher data rate transmissions. The frequency band is generally divided in half, and separated by a central guard band. One half is used for transmit channels (often called "go" channels) and the other half used for receive ("return") channels. A guard band is usually defined at the top and bottom of the band to avoid interference with services operating in adjacent, often adjoining frequency bands. A limited number of transmit and receive frequency channels are defined within each half-band, usually adjacent to one another. Each of these channels has an associated paired channel in the other half of the band. For example, a system might be operating on frequency f1 paired with frequency f1'. The separation of the transmit and receive frequencies is known as the TR spacing, and is usually fixed for a given frequency band. The system can use the lower frequency channels for transmit and the higher frequency channel for receive, or alternatively, it may transmit in the higher channels and receive in the lower channels. Note that in all cases, a limited channel size is defined and the transmission must be kept within this channel. The size of this allocated frequency channel is usually the limiting factor in determining the maximum data rate that the wireless system can carry.

1.3 Spectrum for High Data Rate Wireless Transmission

The key to transmitting high data rate traffic wirelessly is spectrum. As data rate increases, proportionally larger channel bandwidths are required to support the increased throughput capacity. Thus to support the highest data rates, large amounts of spectrum are necessary. In a practical wireless system, the spectrum available for transmission is the channel size allocated by the band manager or regulator. Figure 1.3 shows this trend, plotting commercially available and reported wireless systems data rates versus frequency, showing that increasing carrier frequency does enable increasing data rate. For multi-gigabit per second and higher transmissions, the higher microwave and millimeter-wave frequency bands are required.

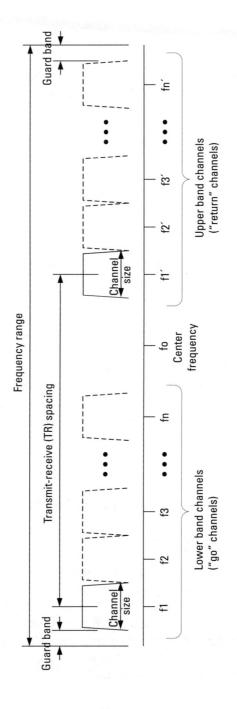

Figure 1.2 Generalized radio frequency band plan.

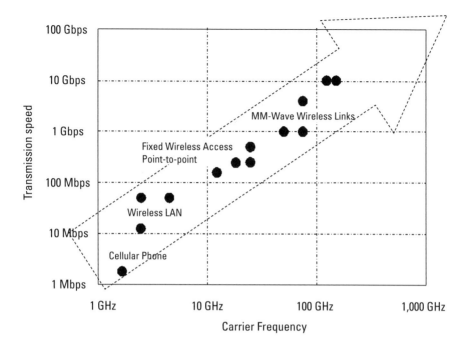

Figure 1.3 To transmit high data rate traffic, high carrier frequencies are required. (*After:* [3].)

Figure 1.4 shows the major frequency allocations available in the United States. Similar allocations, with only minor changes, are in place in most regions of the world. It can be clearly seen that large amounts of spectrum are allocated at the higher frequencies, allowing them to support high data capacities much easier than at lower frequencies. At the microwave bands, limited channel sizes limit data capacities. However there are innovative techniques that allow reuse of the frequencies in these bands, doubling or quadrupling throughput, but at the expense of system complexity. In the millimeter-wave bands at 60 GHz and 70/80 GHz (and to some degree 94 GHz), very wide channels are available, which is ample bandwidth to enable very high data rates to be passed.

1.4 High Data Rate Wireless Technologies

The term "high data rate" is relative. Many different wireless technologies are touted as high data rate, when in fact they are much slower than other common technologies used in different applications. For example, the highest advertised broadband data rates achieved on mobile phones are much slower than high-speed broadband data rates regularly achieved on PCs. Even the concept of broadband is relative. Until recently the FCC in the United States defined

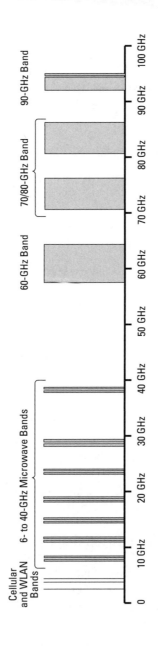

Figure 1.4 Major United States frequency band allocations. (Lower frequency channel widths are drawn overly large for visibility.) (*After:* [3].)

broadband as any service where one-way data rates exceeded 200 kbps. In much of Asia, consumers now consider broadband as approaching speeds of 100 Mbps.

Figure 1.5 shows today's commercially available high data rate wireless technology landscape, showing the comparative data rates achievable under usual operating conditions by several major "high data rate" technologies and the distances over which these data rates can be achieved.

1.4.1 Microwave and Millimeter-Wave Technologies

Only a few technologies are able to reliably deliver gigabit per second and higher data capacity wirelessly. Over short distances, 60-GHz technologies are very applicable. The 60-GHz band covers various frequencies between 57 GHz and 66 GHz in different parts of the world, and is often referred to as "V-band." 60-GHz personal area networks (PAN) systems are available that can deliver data rates of 4 Gbps over distances of about 10m. Specifications are in place to allow data rates beyond 20 Gbps. Longer distance 60-GHz systems can deliver data rates of 1 Gbps over distances of a few hundred meters. Both of these 60-GHz technologies are considered in detail in Chapter 5.

The higher frequency 70/80-GHz band covers a globally available pair of channels from 71–76 GHz paired with 81–86 GHz, which is often referred

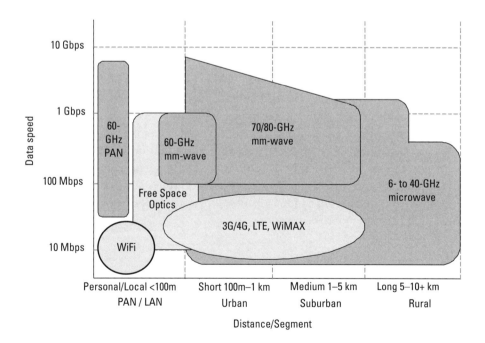

Figure 1.5 The high data rate wireless landscape. (*After:* [3].)

to as "E-band." Because of the large 5-GHz channel sizes, very high data rates can be delivered at longer distances than the 60-GHz band. Systems have been demonstrated that can deliver 6 Gbps, and equipment with data rates of 1 Gbps are widely installed at distances of several kilometers. In the United States there is also a relatively wide allocation at 94 GHz (92–94 GHz and 94.1–95 GHz), which also has a significant bandwidth, but its uneven allocation and proximity to military bands at 94–94.1 GHz make it practically difficult to use. There is also consideration underway to open up further millimeter-wave bands to enable even higher data rate transmissions. There are many unused spectrum blocks allocated for fixed service spectrum in the atmospheric windows around 140 and 240 GHz. By aggregating some of these bands, and opening up wide channels of 50 GHz or higher, regulators could enable 40-Gbps radios with the simplest modulations and 100-Gbps radios with slightly more sophisticated modulation schemes. Chapter 6 covers 70/80 GHz and higher frequency technologies in detail.

Finally 6- to 40-GHz microwave is an established technology that has been used to provide long-distance communications for many years. Transmission of a few hundred megabits per second can be reliability achieved, and these bands are popular for fixed wireless access, including long distance transmission, cellular backhaul (connecting together cell sites), and enterprise connectivity (connecting together buildings and campuses). National regulators have deliberately sliced the available microwave spectrum into many narrow channels to encourage competition and permit licensees to use the services without interference. Typically these channels are up to 50 MHz in the United States and up to 56 MHz in the rest of the world. Since these bands are licensed, operators own the frequencies and are guaranteed interference-free transmissions and so can use technologies optimized for data throughput. Despite being limited by relatively small channel sizes, sophisticated signal processing techniques and complex modulation schemes allow systems to squeeze data rates of about 300 to 350 Mbps in these narrow channels. Furthermore, parallel links can be employed with techniques that efficiently reuse transmission frequencies for dual or quadruple data streams, increasing the effective data throughput to beyond 1 Gbps over short and medium distances. Chapter 4 provides a detailed examination of such high data rate microwave systems.

1.4.2 Free-Space Optics

Another wireless technology capable of achieving multi-gigabit wireless transmission is free-space optics (FSO). FSO technologies operate in the very highest regions of the frequency spectrum, near visible light. FSO technologies employ a laser transmitter to generate a focused optical light wave that carries data through the atmosphere to an optical receiver located at a fixed distance from the transmitter. Very wide bandwidths are available in the optical part of the spectrum that enables very high data rate transmission. Commercial systems are available that operate to 1 Gbps and experimental links have been demonstrated at up to 40 Gbps. Because FSO systems operate at unregulated frequencies where there are no licensing restrictions, links can be installed and commissioned quickly and with minimum hurdles.

Despite much excitement when first commercially introduced, FSO links have not gained traction in the high data rate marketplace because of numerous practical and operational difficulties associated with the physics of laser transmissions. Communications at optical frequencies are especially sensitive to atmospheric and environmental conditions, particularly fog, which can drastically limit distance. A thick fog with 0.1 g/m^3 vapor density (about 50m visibility) will result in 225 dB/km attenuation at FSO frequencies. (The same fog gives just 0.4 dB/km attenuation at 70/80 GHz.) Similar link degradation occurs with any airborne particles such as snow, sand, dust, flying debris, or even residue from agricultural burning. Since fog occurrences are difficult to predict, and when experienced can last for many hours, FSO is not a practical technology for high-availability, carrier-grade equipment, or for systems operating beyond short distances in fog-prone areas.

A number of physical effects also need to be accounted for and overcome in designing, planning, and installing any high-resilience, single-beam optical transmission path:

- *Pointing effects:* Birds flying through a narrow optical beam can block the path, causing momentary outages that would affect timing-sensitive data traffic.

- *Precise alignment:* Tower sway or movements of solid buildings as they naturally heat up and cool down during the day can misalign narrow beam systems.

- *Scintillation:* Longer east/west-facing optical links can be affected by diurnal sunlight effects.

- *Laser aging and safety:* The usable life of any laser deteriorates through laser use and over time. The use of laser systems also raises concerns over eye safety.

To compensate for these effects, higher quality FSO equipment employs multiple-beam architectures and beam-tracking technologies. Laser heating and cooling is used to maintain lasers within optimal temperature ranges to limit aging. These equipment enhancements result in complex equipment, with increased cost and reduced equipment reliability consequences. For these reasons, FSO is not a commercially viable, mass market high data rate technology.

However it is being used for a number of lower data rate, short-distance applications. Since FSO has only very limited use for commercially viable high data rate transmission, it is not considered further in this text.

1.4.3 WiFi, WiMAX, and LTE

In the popular literature, much has been written about high data rate multiple-access Wireless Fidelity (WiFi), Worldwide Interoperability for Microwave Access (WiMAX), and Long Term Evolution (LTE) wireless systems for consumer applications, which operate in the frequency bands below 6 GHz. As discussed earlier, the term high data rate is very subjective, and such residential systems operate with much slower data speeds than industrial systems operating at frequencies of 6 GHz or higher.

Practically realizable data rates of the WiFi, WiMAX, and LTE technologies, despite marketing hype to the contrary, are much lower than advertised, hampered by the shared wireless protocols employed and the fact that bandwidths allocated to such services are usually not large enough to support the higher advertised data rates. This is depicted in Figure 1.6, which shows theoretical and practically realizable data rates for a variety of technologies available today.

Consider WiFi, for example. WiFi is a short distance, multiaccess technology operating in the 2.4- and 5.8-GHz unlicensed bands. WiFi is popular for local area residential or "hot spot" connectivity. The widely used IEEE 802.11g version allows theoretical data rates to 54 Mbps, and the most recently published variation of the standard, IEEE 802.11n-2009, offers theoretical peak data rates up to 600 Mbps. This latter technology employs 4 spatial streams, multiple antennas, complex beamforming, and bonded channel transmissions to reach such theoretical speeds. In practice, lab experiments have yielded data rates to 200 Mbps, but practical user speeds are much less. The reason for this is, like all

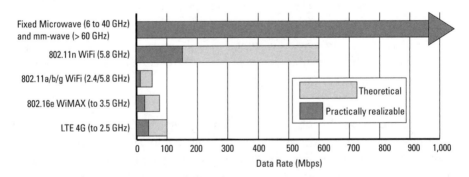

Figure 1.6 Theoretical peak data rates and practically realizable data rates for available commercial technologies.

multiple-access technologies, WiFi has a number of limitations that significantly reduce practical speeds below theoretical peaks and laboratory demonstrations. In real life, data rates are dependant on the environment: that is, the distance from the access point, the number of users sharing the capacity, the number of channels available to be bonded, and the usually constrained access point's broadband connection. For this reason, users typically realize 1 to 10 Mbps connectivity in commercial hot spot environments; much less than the theoretical maximums. In addition, by necessity WiFi is an unlicensed, broadcast technology, resulting in interference, data contention, and security concerns for installers and system architects.

Fourth-generation (4G) wireless systems promise a substantial increase in broadband wireless data throughput over existing second and third generation (3G) cellular systems. 4G is a broad term that has unfortunately been misapplied and misused. Per the ITU definition, a 4G system has a peak data rate of 100 Mbps under high mobility conditions and 1 Gbps for low mobility and stationary conditions. The Third Generation Partnership Project (3GPP) Long Term Evolution (LTE) technology is often branded as 4G, but is limited to theoretical bit rates of around 100 Mbps under stationary conditions. WiMAX is also often branded as a 4G technology, but available networks have capacities that fall far short of this. Strictly speaking, current implementations of both LTE and WiMAX are more accurately described as 3.5G systems.

Operating primarily in the 2 GHz and 3.5 GHz licensed bands, a number of WiMAX networks and LTE trials are in place around the world. WiMAX addresses many of the quality-of-service (QoS) and security issues inherent with WiFi, and is usually implemented using secure licensed frequency bands. Theoretical peak data rates of close to 100 Mbps are possible, although real implementations provide sustained data rates of less than 10 Mbps. Future extensions to the WiMAX family (for example IEEE 802.16m) will further extend user data speeds and experiences. WiMAX does offer the benefit of mobility, making the analogy to advanced cellular systems more accurate than to WiFi networks. LTE by contrast has been designed as the next generation of the existing 3G cellular technologies. Theoretically, data rates beyond 100 Mbps are possible when the full LTE-Advanced standards are implemented. However, current releases offer data rates higher than WiMAX networks, with practically realized sustained data rates of a few tens of megabits per second.

1.5 Summary

While fiber remains the transmission medium of choice for high data rate networking and multi-gigabit data transmission, it has a number of limitations. It is expensive and time consuming to install, commission, and/or lease, and it is not widely available to commercial buildings. An opportunity exists for wireless technologies that are able to provide cost-effective wireless gigabit per second transmission capabilities, to provide an alternative to buried fiber.

For high data rate wireless transmissions, wide channel sizes are required. Fortunately, band managers and frequency regulators have recognized this and allocated some higher frequency bands with very wide channel bandwidths, allowing very high data rate communications to be supported. The 60-GHz and 70/80-GHz frequency bands in particular have very wide frequency allocations, and are able to support multi-gigabit per second data rates. Even though use of these bands is in its infancy, both 60 GHz and 70/80 GHz commercial equipment has been available for several years that can offer transmission speeds in excess of 1 Gbps. Also of interest for gigabit transmissions are the lower frequency 6- to 40-GHz microwave bands. Although these have relatively narrow channel sizes, sophisticated signal processing technologies and frequency reuse techniques allow vendors to offer systems that exceed 1 Gbps transmission speeds.

Both microwave and millimeter-wave technologies have relative strengths and weaknesses, indicating that each has a role to play depending on the end user's different installation needs, link performance requirements, and budget. However both are able to offer speeds in excess of 1 Gbps, and as such, are able to challenge fiber for ultrahigh speed data transmission.

References

[1] ITU-R Rec. V.431-7, "Nomenclature of the Frequency and Wavelength Bands Used in Telecommunications," 2000.

[2] IEEE 521-2002, "IEEE Standard Letter Designations for Radar-Frequency Bands," 2003.

[3] Wells, J.A., "Faster Than Fiber: The Future of Multi-Gb/s Wireless," *IEEE Microwave Magazine*, Vol. 10, No. 3, 2009, pp. 104–112.

2

High-Frequency Wireless Propagation

2.1 Introduction

For a proper understanding of any wireless system and the design of any wireless link, a detailed knowledge of wireless propagation is necessary. Wireless signals are electromagnetic waves that follow classic Maxwell's equations. If such a wave is propagated in free space, the path followed will be a straight line and can be considered as an infinitely wide, infinitely long series of wave fronts. However, in reality, a radio wave propagating through the Earth's atmosphere will encounter many atmospheric effects, which will hinder transmission. For high-frequency radio systems, the atmosphere is the most significant factor in limiting wireless performance, causing outages or deterioration in wireless transmissions. The principal characteristic of the atmosphere that causes this limitation is attenuation. There are regions of the high-frequency spectrum that are attenuated by as much as several hundred dB/km, which makes operation at ranges of even a few tens of meters extremely challenging. There are other "atmospheric windows" where attenuation is significantly below 1 dB/km, which makes long-range communications entirely practical.

The most significant atmospheric attenuation effects that need to be considered for high-frequency wireless systems are as follows:

- Free space loss;

- Attenuation due to atmospheric gases;
- Attenuation due to precipitation (rain);
- Attenuation due to other particles in the atmosphere (e.g., fog, snow, ice, dust, and sand);
- Attenuation due to interference of the signal (e.g., foliage or other items partially or fully blocking the transmission path).

This chapter presents an introduction to electromagnetic wireless propagation, and then a detailed discussion of each of these propagation effects and their individual characteristics, so that proper path planning and link budgets can be developed.

It should be noted that the effects discussed in this chapter are applicable to all wireless systems of a few gigahertz and above. Lower-frequency transmissions have some unique effects, which are not considered here, since these lower-frequency bands do not have the ability to support gigabit per second communications. Note too that this chapter is equally applicable to terrestrial, ground to air, high altitude, and even satellite systems, where the same theories and concepts are equally applicable and all hold true.

2.2 Electromagnetic Propagation

A radio signal travels through free space as a spatially and time-varying electric and magnetic field. Both this electric and magnetic field are dependent on the other. An oscillating electric field generates an oscillating magnetic field, which in turn generates an oscillating electric field, and so on. Together these oscillating fields form an electromagnetic wave. Electromagnetic propagation is fully described by Maxwell's equations, which are well covered in the literature (for example, [1]). This section introduces the basic concepts of relevance to high-frequency propagation.

Figure 2.1 shows the form of an electromagnetic wave as it propagates through a medium. A sinusoidal electric field E is at right angles (said to be

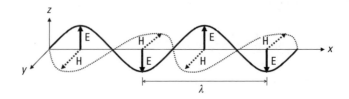

Figure 2.1 An electromagnetic wave at an instant of time.

orthogonal) to a similar sinusoid magnetic field H, both of which are orthogonal to the direction of travel. This waveform is known as a transverse electromagnetic (TEM) wave.

An electromagnetic wave has four fundamental properties: frequency, amplitude, phase, and polarization. As shown in later chapters, these four properties are important because varying these allows additional communication information to be carried on the electromagnetic wave. The first three fundamental properties, frequency, amplitude and phase, are related by (2.1), which describes the electric E-field propagation:

$$E = E_0 \cos(\omega t + \phi) \qquad (2.1)$$

where E_0 is the amplitude, ω is the angular frequency, and ϕ is the phase of the time-varying sinusoidal signal. The magnetic H-field can be similarly described. The angular frequency ω, measured in radians, is more normally described as a frequency f, measured in hertz (Hz), whereby:

$$\omega = 2\pi f \qquad (2.2)$$

Electromagnetic waves are often described by their wavelength λ. As shown in Figure 2.1, this is the distance between two points on the sinusoidal waveform having the same phase. Wavelength is dependent on the frequency of the signal and the speed with which it is travelling, and can be described by:

$$\lambda = \frac{v}{f} \qquad (2.3)$$

where v is the velocity of propagation of the wave, measured in m/s. This is dependent on the medium through which the electromagnetic wave is passing, which is given by:

$$v = \frac{1}{\sqrt{\varepsilon \mu}} \qquad (2.4)$$

where ε and μ are the permittivity and permeability of the material, respectively. In a vacuum these parameters are described by the permittivity of free space ε_0 equal to 8.854×10^{-12} F/m, and the permeability of free space μ_0 equal to $4\pi \times 10^{-7}$ H/m. Thus,

$$v_{vacuum} = \frac{1}{\sqrt{\varepsilon_0 \mu_0}} = 3 \times 10^8 \ \text{m/s} \tag{2.5}$$

which is equal to c, the speed of light. Therefore, in a vacuum, electromagnetic waves travel at the speed of light. When travelling through other mediums, the values of ε and μ are increased relative to the capacitance or inductance of that medium verses a vacuum. This means that the dielectric properties of the transmission medium will always slow down a propagating signal to less than the speed of light. Conveniently, the permittivity and permeability of air are approximately the same as in a vacuum, and so electromagnetic transmissions in free space are considered to travel at the speed of light.

Analogous to electronic signals, every transmission medium has an impedance, dependent on the permittivity and permeability of the medium. This impedance impairs an electromagnetic wave's propagation. In free space, this impedance is described by:

$$Z_o = \frac{E}{H} = \sqrt{\frac{\mu_0}{\varepsilon_0}} = 377\Omega \tag{2.6}$$

where Z_o is known as the impendence of free space and is equal to 377 ohms.

Polarization of an electromagnetic wave refers to the plane in which the signal's electric field propagates. In Figure 2.1, the E-field is seen to propagate in a vertical plane, and so the signal is said to be vertically polarized. Electromagnetic signals can be vertically polarized, horizontally polarized, or polarization can be set at any arbitrary angle. It is even possible to produce circular polarized signals by continuously rotating the plane of polarization.

2.3 Free-Space Loss

Free-space loss is the dominant form of loss in a wireless link. This is the physical loss associated in transmitting from point A to point B.

Free-space loss is derived by considering the energy emitted from a radiation source. Consider an isotropic point source A, which by definition, radiates uniformly in all directions (with unity gain or a gain of 0 dB). Let the total power radiating from this point transmitter source be P_T. At any one time, the radio waves propagating from this point source will form a wave-front travelling outwards from the point source in a sphere of surface area $4\pi R^2$, where R is the radius of the sphere. At any point on the surface of this sphere, the power density (power flow per unit area, or flux) is therefore given by:

$$P_{density} = \frac{P_T}{4\pi R^2} \qquad (2.7)$$

It can be seen that the power density is proportional to $1/R^2$, meaning that the power per unit area drops quickly as the radio wave moves away from the transmitter.

Now consider a receiving antenna at point B on the sphere's surface. Conventionally, an isotropic antenna may function as either a transmitting antenna or a receiving antenna. As a receiver, it absorbs power from the radiation field in which it is situated. The amount of power that a receiver absorbs in relation to the power density of the field in which it is placed is determined by the receiving antenna's effective aperture. For an isotropic antenna, the effective area is $\lambda^2/4\pi$, where λ is the wavelength of the incident wave front. Therefore, from (2.7) it follows that an isotropic receiver, situated in a radiation field, will have an incident power P_R equal to:

$$P_R = P_T \left(\frac{\lambda}{4\pi R} \right)^2 \qquad (2.8)$$

Equation (2.8) is commonly known as the Friis equation. It shows that in a radiation field, the power received by an isotropic receiver is equal to the power emitted by an isotropic transmitter, reduced by an amount equal to $(\lambda/4\pi R)^2$. This reduction factor, due only to the physical separation of the transmitter and receiver elements A and B, is known as the free-space loss.

For a practical wireless system, it is more convenient to replace the ideal isotropic transmitter P_T with a real transmitter of power P_{tx} and gain G_{tx}, and the ideal isotropic receiver P_R with a real receiver of power P_{rx} and gain G_{rx}. Substituting into (2.8), converting in logarithms, and rearranging gives the following:

$$P_{rx}(\text{dBm}) = P_{tx}(\text{dBm}) + G_{tx}(\text{dB}) + G_{rx}(\text{dB}) - 20\log_{10}\left(\frac{4\pi R}{\lambda} \right) \qquad (2.9)$$

Equation (2.9) describes the receiver power detected in a full radio path. It includes the transmitter power, both the transmitter and receiver antenna gains, and the free-space loss due to the physical separation of the transmitter and receiver. This free-space loss, L, is dependant only on the frequency of the transmission and the distance between the antennas, and is given by:

$$L(\text{dB}) = 20\log_{10}\left(\frac{4\pi R}{\lambda}\right) \qquad (2.10)$$

A more suitable equation for describing free-space loss is derived by converting (2.10) to use more convenient units of measurement:

$$L(\text{dB}) = 92.4 + 20\log_{10} D(\text{km}) + 20\log_{10} f(\text{GHz}) \qquad (2.11)$$

where D is the antenna separation (or link distance) measured in kilometers and f is the transmission frequency measured in gigahertz.

Figure 2.2 shows different values of free-space loss over a range of common high transmission frequencies and distances. It is clear that even for very short distances, free-space loss can be significant. For example, at 60 GHz, even a 10-m link experiences 88 dB of free-space loss. Typical terrestrial links of 1 to 10 km exhibit 120 dB to 150 dB, depending on actual frequency and distance. The graph shows that for every octave increase (or doubling) in range, atmospheric attenuation increases by 6 dB, and that for every decade (or 10 times) increase in range, attenuation increase by 20 dB. Since free-space loss is proportional to the square of the link distance, and long distance (many kilometers distance) terrestrial links are common, the possibility of high-frequency satellite or ground

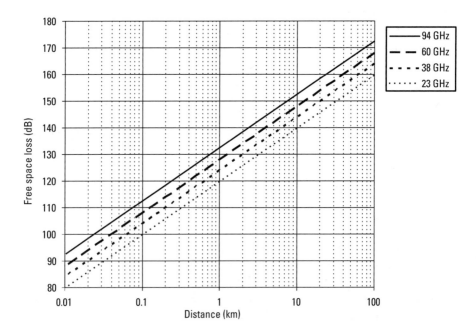

Figure 2.2 Free-space loss verses distance for various common frequencies of operation.

to air communications are quite possible. High altitude and other space-born applications may experience a free-space loss only a few dBs or so more than the longer terrestrial links.

2.4 Attenuation Due to Atmospheric Gases

In addition to free-space loss, millimeter-wave transmissions experience further attenuation as millimeter-waves are absorbed by molecules of oxygen, water vapor, and other gaseous constituents in the atmosphere. At some frequencies, atmospheric absorption is very low, meaning that long-distance communications can be readily achieved. This is useful for terrestrial trunking systems or ground to air applications. At other frequencies, atmospheric attenuation can be very significant, limiting transmissions to short distances. This can be exploited for secure communications where eavesdropping beyond the intended communication path is minimized due to rapid deterioration of the transmitted signal strength. Similarly, systems employing retransmission can utilize this signal attenuation to enable more efficient frequency reuse. For very long propagation paths, for example ground to air satellite or other high-altitude platform systems (HAPS), understanding of atmospheric absorption is important.

Atmospheric loss varies significantly with frequency, due to the mechanical resonant frequencies of different gas molecules that occur naturally in the atmosphere. In the frequency region up to 200 GHz, for example, water vapor (H_2O) absorption arises from a weak electric dipole rotational transition at 22.235 GHz and a much stronger transition at 183.31 GHz. Similarly, oxygen (O_2) absorption occurs at around 60 GHz because of a multitude of magnetic dipole transitions centered close to this frequency, and also at the single frequency of 118.75 GHz. Although quantum mechanics suggests discrete resonant lines and therefore precise absorption frequencies, molecular collisions within these transitions causes pressure broadening, expanding the frequency characteristics of the absorption peak beyond the single frequency resonant lines. There are many other atmospheric gases and pollutants that have absorption lines in the high frequency bands (for example SO_2, NO_2, and N_2O); however due to their low density, their absorption losses are negligible. Water vapor and oxygen are the principal absorption constituents in the atmosphere.

Detailed calculations of atmospheric absorption by water vapor and oxygen were first published in the 1940s. These initial quantum mechanic calculations were later refined and improved using laboratory measurements to yield several absorption models that are widely used today in the propagation and remote-sensing communities. A detailed account of these methodologies and history is given in [2]. All the models are strongly dependent on environmental conditions such as temperature, pressure, and humidity, which in turn are dependent on such factors as weather patterns, geographical region, and elevation.

One method widely used in the wireless communications community is that published by the ITU [3], which characterizes atmospheric absorption at frequencies up to 1,000 GHz. The ITU defines atmospheric attenuation γ for terrestrial paths close to the ground as:

$$\gamma = \gamma_0 + \gamma_w \tag{2.12}$$

where γ_0 is the specific attenuation due to dry air (predominantly oxygen and nitrogen) and γ_w is the specific attenuation due to water vapor. Each of these are evaluated as functions of pressure, temperature, and humidity by summing individual molecular resonance lines from oxygen and water vapor, plus some small additional factors to account for second-order physical effects. Thus (2.12) becomes:

$$\gamma = \gamma_0 + \gamma_w = 0.182 f \left(\sum_i S_i F_i + N(f) \right) \tag{2.13}$$

where f is frequency measured in GHz, S_i is the strength of the i-th line, and F_i is the line shape factor summed over all the oxygen and water vapor resonant lines, and $N(f)$ is a final frequency-dependent factor to account for nitrogen absorption. Reference [3] further characterizes the strength and line shape expressions, and how they vary with temperature, pressure, and humidity, for each of the 44 unique oxygen and 34 unique water vapor resonances below 1,000 GHz.

Figure 2.3 plots (2.13) showing atmospheric absorption up to 1,000 GHz. The major absorption resonance lines are clearly visible. Plots for both standard atmospheric conditions (15°C temperature, 1,013-hPa pressure, 7.5-g/m³ humidity) and for dry air (the same temperature and pressure, but with no water vapor or 0% humidity) are shown. The strong dependence on water vapor is evident. The water vapor peaks at 22.235 GHz and 183.31 GHz are clearly visible, as are the broad oxygen peak at 60 GHz and the additional oxygen absorption line at 118.75 GHz. Many more molecular resonance peaks are seen at frequencies higher than 300 GHz. Radiometric studies and remote temperature and water vapor sensing have long been undertaken at the absorption peaks at around 60, 118, 183 GHz and higher.

Wireless spectrum regulators have paid close attention to these attenuation characteristics when developing wireless communication rules and regulations. For example, atmospheric attenuation is negligible at frequencies up to about 50 GHz. These frequencies are widely available worldwide for a variety of

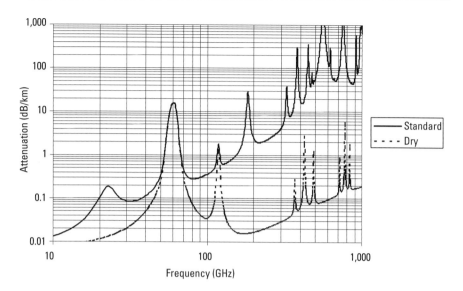

Figure 2.3 Atmospheric absorption of standard and dry air at sea level. (Conditions: 15°C temperature, 1,013-hPa pressure, with 7.5-g/m³ humidity for standard atmospheric conditions and zero humidity for dry air.)

wireless applications and communication systems. The strong oxygen absorption extending from about 50 to 70 GHz yields at its peak about 15 dB/km attenuation, making this frequency range unsuitable for long-distance communications. For this reason, wide swaths of frequencies at around 60 GHz are available in many parts of the world for primarily short distance, unlicensed wireless use. The atmospheric minima at 94 GHz (for dry air) has long been used by military and government agencies, where the low atmospheric attenuation and general lack of services means interference-free covert operations can be undertaken. In addition, the low attenuation window from around 70 to 100 GHz where atmospheric absorption is less than 0.5 dB/km has recently been opened up for services such as ultrahigh data rate communications in the 71- to 86-GHz band, and automotive radar and vehicular collision avoidance systems at the 77-GHz band. Finally, the wide atmospheric windows at 140 GHz and 240 GHz where relatively low attenuation values are still observed, are being considered for very high data rates systems for the future [4].

2.4.1 Regional Variations

Atmospheric attenuation has a strong dependence on environmental conditions; particularly temperature, pressure, and humidity (the amount of water vapor in the air). For convenience, the ITU defines a reference standard atmosphere as follows:

- Temperature: 288.15K (15°C)
- Pressure: 1,013.25 hPa (1 atmosphere)
- Water vapor: 7.5 g/m^3

However, these parameters vary strongly with regional metrological conditions that differ significantly in different parts of the world. Global temperature varies considerably, as does humidity. The amount of water vapor in the atmosphere at sea level can vary from 0.001 g/m^3 in a cold, dry climate to as much as 30 g/m^3 in hot, humid climates. To provide some granularity, the ITU has characterized a set of geographical profiles to provide more regional characterization and closer approximations of localized temperature, pressure, and humidity profiles [5]. This defines three regional zones, described below and shown diagrammatically in Figure 2.4:

- *Low latitude:* The equatorial zone between ±22°, including Central America, northern South America, central Africa, southern Asia, and northern Australia.
- *Mid latitude:* The region between 22° and 45°, both north and south of the equator, including the continental United States, northern Africa, South Africa, southern Europe, southern South America, central Asia, and southern Australasia.
- *High latitude:* The regions above and below 45° north and south, including Canada, Alaska, central and northern Europe, Russia, and Antarctica.

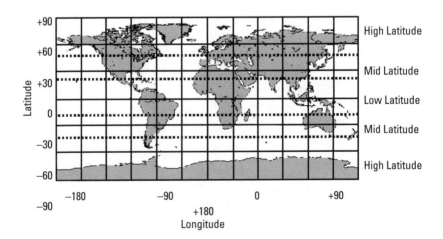

Figure 2.4 ITU-defined regional zones.

Furthermore, the ITU considers seasonal variations within these zones. For low latitudes, the seasonal variations are not very significant and a single annual profile is defined. For mid and high latitudes, more seasonal variations are seen and both summer and winter profiles are derived. Table 2.1 shows the ITU's reference atmospheric profiles for each of these regional profiles.

Knowing these regional profiles, localized values for temperature, pressure, and humidity can be used in (2.13) to provide more localized predictions of atmospheric attenuation characteristics. Figure 2.5 shows atmospheric absorption at up to 120 GHz for each of these regions, showing the strong dependence on environmental conditions and widely varying atmospheric attenuations that can occur. For example, in the continental United States (mid lat), atmospheric attenuation in the 71–86-GHz terrestrial band is typically 0.5 dB/km in the summer, dropping to 0.3 dB/km in the winter. In Northern Europe (high lat) much lower levels of atmospheric absorption—below 0.1 dB/km—would be expected. Table 2.2 provides actual attenuation values at typical high data rate communication frequencies.

2.4.2 Altitude Variations

Since atmospheric absorption is strongly influenced by temperature, pressure, and humidity, atmospheric absorption also varies significantly with altitude. Propagation-wise, high-altitude communications using high-frequency microwaves or millimeter-waves are very favorable. Thus millimeter-waves are especially useful for HAPS such as satellite to satellite and high-altitude to ground applications.

To facilitate above sea level modeling of atmospheric absorption, the ITU has published guidelines for standard atmospheric condition variations with altitude [5]. This model divides the atmosphere into seven successive layers, each showing a linear variation with temperature. Resulting pressures are derived, up

Table 2.1
Reference Metrological Regional Profiles

Profile	Latitude ($\pm°$)	Temp (°C)	Pressure (hPa)	Water vapor (g/m³)
Reference standard atmosphere	n/a	15	1,013	7.5
Low latitude	< 22	27.4	1,012	19.7
Mid latitude, summer	22–45	21.8	1,013	14.4
Mid latitude, winter	22–45	−0.4	1,019	3.5
High latitude, summer	> 45	13.7	1,008	9.0
High latitude, winter	> 45	−15.7	1,011	1.2

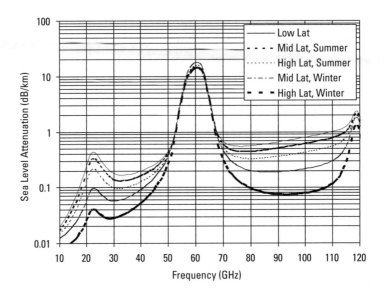

Figure 2.5 Atmospheric absorption for five regional and seasonal profiles.

Table 2.2
Atmospheric Attenuation Values at Typical High Data Rate
Communication Frequencies for Five regional and seasonal profiles.

| Profile | Sea Level Atmospheric Attenuation (dB/km) | | | | | |
	23 GHz	38 GHz	60 GHz	73 GHz	83 GHz	94 GHz
Low Lat	0.438	0.185	13.8	0.571	0.583	0.712
Mid Lat, Summer	0.330	0.149	14.5	0.478	0.460	0.552
Mid Lat, Winter	0.095	0.078	17.7	0.294	0.189	0.196
High Lat, Summer	0.217	0.113	15.5	0.382	0.328	0.381
High Lat, Winter	0.040	0.044	13.9	0.176	0.084	0.074

to an altitude of 85 km, beyond which the thermodynamic equilibrium of the atmosphere starts to break down, and the hydrostatic equation, on which the above equations are based, is no longer valid. The distribution of water vapor in the atmosphere is generally highly variable, and is approximated by the model. Variations of temperature, pressure, and humidity per this ITU model are depicted in Figure 2.6.

To facilitate more localized modeling, reference [5] also details two other models. The first includes monthly averages of vertical temperature, pressure, and relative humidity profiles calculated for 353 locations over the world, using 10 years of meteorological observations. This data includes noon and midnight

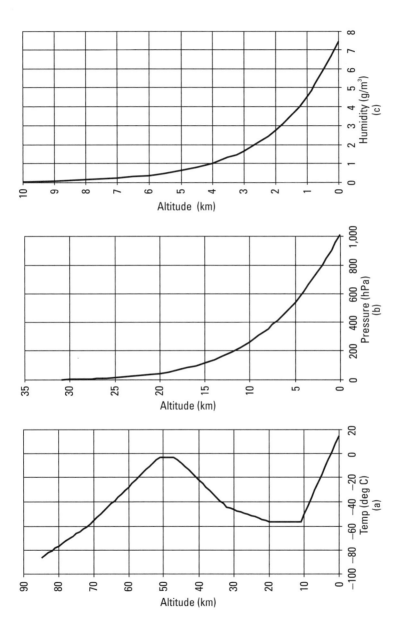

Figure 2.6 Variations in standard temperature (a), pressure (b), and humidity (c) with altitude.

profiles calculated in the absence of rain, from 0 to 16 km altitude. The second model includes similar monthly averages of vertical temperature, pressure, and water vapor profiles up to 30-km above the Earth's surface, but this time for 6 hourly intervals. In this second model, 15 years of data was used, for locations from 0 to 360 in longitude and from +90 to –90 in latitude, with a resolution of 1.5° in both latitude and longitude.

Considering the standard atmospheric models shown in Figure 2.6, (2.13) can be used to derive variations in atmospheric attenuation over altitude. This is depicted in Figure 2.7, showing that attenuation at all frequencies except the molecular resonances drops significantly with height. This is one reason why the higher-frequency bands are attractive for high data rate high-altitude systems.

Atmospheric attenuation at the 60-GHz oxygen absorption peak is worthy of more attention, since this frequency plays an important role in short distance, high data rate communications. All the previous figures have shown this as a single broad peak extending from about 50 to 70 GHz. In fact, there are 37 unique oxygen absorption lines, which at high pressure merge together to form a single, broad absorption band, peaking at about 15 dB/km attenuation at sea level. Figure 2.8 shows the effect of altitude on this 60-GHz resonance, where the reduction in pressure with height reveals these individual resonance lines.

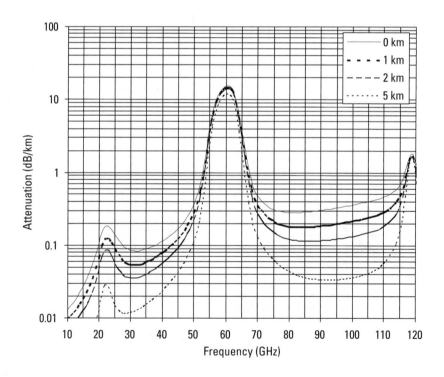

Figure 2.7 Variations in atmospheric attenuation with altitude.

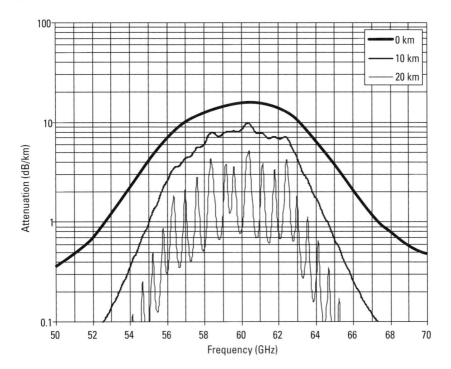

Figure 2.8 Characterization of atmospheric attenuation at the 60-GHz oxygen absorption peak with altitude.

This spiky attenuation characteristic is a significant impediment to 60 GHz broadband communications at high altitudes, where costly and complex signal equalization techniques would be required for receivers to properly demodulate wideband signals. Despite the reduction in overall attenuation, these 60-GHz altitude characteristics make 60 GHz a poor choice for high-altitude systems.

2.5 Attenuation Due to Rain

In systems operating to about 10 GHz, transmission loss is primarily accounted for by the free-space loss. However, in the microwave bands above 10 GHz and the higher millimeter-wave bands, it has already been shown that atmospheric absorption needs to be considered. At these higher frequencies, rain attenuation also needs to be accounted for.

The attenuation due to rainfall is strongly dependent on the size and distribution of the water droplets. Raindrop-size distribution depends strongly on rain rate. Larger drops occur at higher rain rates. Large raindrops can be a few millimeters in diameter, corresponding to free-space wavelengths of about 100

GHz or so. Thus electromagnetic scattering from rain starts to be observed at frequencies of about one-tenth of this, or about 10 GHz and above.

If a raindrop is considered as spherical, Mie theory, which describes the scattering of electromagnetic waves from a spherical particle, can be used to calculate rain attenuation. In reality, however, falling raindrops are not spherical, as air friction flattens the water droplet as is falls through the air. Since the vertical dimension of falling raindrops are therefore less than the horizontal dimensions, rain attenuation for vertically polarized wireless signals will always be less than for horizontally polarized waves at a fixed frequency. Furthermore, since raindrops are composed of water, the dielectric constant of liquid water needs to be accounted for in attenuation analyses.

The ITU has studied rain and its effects on electromagnetic propagation in detail and developed a methodology for characterizing rain attenuation [6]. This model describes the atmospheric attenuation due to rain γ_R (measured in dB/km) as a function of rain rate R (measured in mm/hr) using the following equation:

$$\gamma_R = kR^\alpha \tag{2.14}$$

where k and α are coefficients determined from curve fitting to experimental data. Both k and α are dependent on polarization. Values for both k and α, for both horizontal and vertical polarizations, for frequencies from 1 GHz to 1,000 GHz are shown in Figure 2.9.

It can be seen that at the millimeter-wave frequencies, there is very little difference between the horizontally polarized and vertically polarized coefficients, showing that for high-frequency communication systems, no difference in attenuation would be expected between vertical and horizontal polarized transmissions. This is not the case at microwave frequencies (40 GHz and below) where traditional wisdom is that vertical polarization should always be the default polarization due to better rain resilience. This is confirmed by Figure 2.9, whereby a noticeable deviation in the polarization coefficients can be observed at the lower microwave frequencies.

The ITU model also shows how these coefficients can be modified for circular polarization and also for angled (slant) elevation links.

Using the coefficients in Figure 2.9 and (2.14), rain attenuation for different rain rates and frequencies can be derived. This is shown in Figure 2.10 for vertically polarized signals. This graph confirms that at 10 GHz and below, except in the most intense cases, rainfall has negligible effect on signal attenuation. The graph also shows that at certain frequencies, rain attenuation peaks. This is due to a Mie resonance occurring in this region, corresponding to maximum Mie scattering. At higher frequencies, after the Mie resonance has been exceeded, the attenuation curves flatten off as Mie scattering remains almost constant well into the optical visible frequency range.

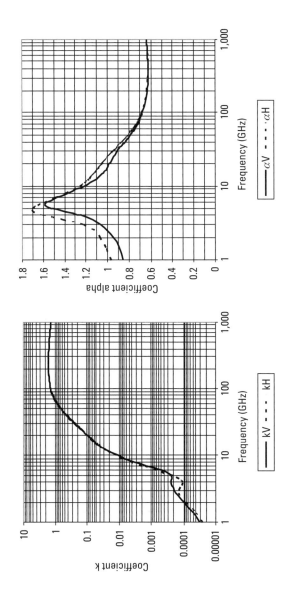

Figure 2.9 Values for *k* and *α*, for both horizontal and vertical polarizations, for calculating rain attenuation at up to 1,000 GHz.

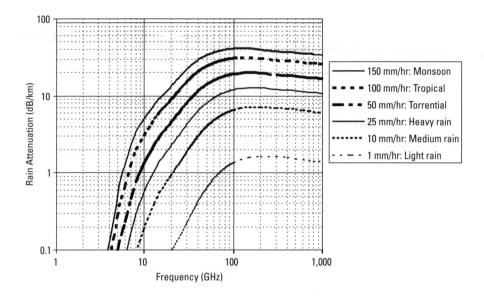

Figure 2.10 Rain attenuation versus frequency for various intensity rain rates.

Figure 2.10 demonstrates that at higher frequencies, rain can cause significant signal deterioration. At 150 mm/hr (6 in/hr) monsoon rain rates, for example, some links can be attenuated up to 40 dB/km. Fortunately rainfall intensity at this rate is very rare and localized to just a few regions of the world. However rainfall rates at 100 mm/hr (4 in/hr) are experienced in the United States and Europe, which can yield attenuations to about 30 dB/km in the commercially available 71–86-GHz millimeter-wave bands. Figure 2.11 shows attenuations versus rain rates for commonly used high-capacity frequency bands. Chapter 7 provides much more detail of rainfall and rain intensity distribution, and how they vary globally. This allows regional link planning, allowing calculations of link distances, link availability (uptime), and outages (downtime) to be calculated, and tradeoffs between them established.

2.6   Attenuation Due to Other Airborne Particles

2.6.1   Fog and Clouds

Both fog and clouds consist of small water droplets, generally of a size less than 0.1 mm. This dimension is equivalent to a free-space wavelength of 3,000 GHz or more. Mie scattering from such water particles would therefore be expected for submillimeter-wave frequency systems operating at around this frequency. For transmissions up to these frequencies, where the particles are sufficiently

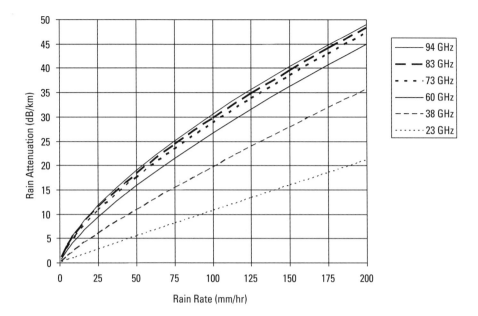

Figure 2.11 Rain attenuation versus rain rate for major high data rate communication frequencies.

small compared to the wavelength, Rayleigh scattering should be considered. However at frequencies below approximately 100 GHz, this attenuation is small.

Similar to other weather parameter characterizations, the ITU has produced a model for electromagnetic radio attenuation from both fog and clouds [7]. This model shows that specific attenuation due to fog and cloud water droplets can be written in terms of total water content per unit volume:

$$\gamma_C = K_l M \tag{2.15}$$

where γ_C is the specific attenuation within the fog or cloud (measured in dB/km), K_l is the specific attenuation coefficient [measured in (dB/km)/(g/m³)], and M is the liquid water density in the cloud or fog (measured in g/m³). The liquid water density M in fog is typically about 0.05 g/m³ for medium fog (visibility of the order of 300m) and 0.5 g/m³ for thick fog (visibility of the order of 50m).

Reference [7] provides details of the specific attenuation coefficient K_l, and shows that it is dependent on both the frequency and the dielectric constant of water (which in turn is dependant on both frequency and temperature). A plot of this value of K is shown in Figure 2.12 for frequencies up to 1,000 GHz and for a range of temperatures found in fog and clouds.

Using the coefficients of K_l from Figure 2.12, (2.15) can be used to determine the attenuation due to fog and clouds. Figure 2.13 shows calculated fog attenuation over frequency for both medium fog (0.05 g/m³ humidity) and

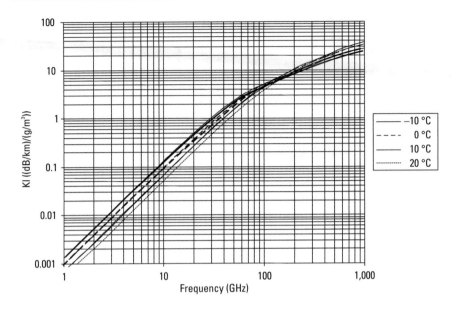

Figure 2.12 Frequency and temperature dependence of the specific attenuation coefficient K_l.

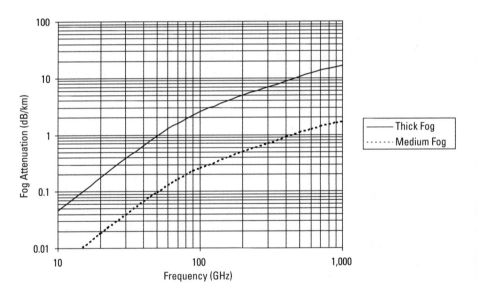

Figure 2.13 Fog attenuation at 0°C for medium (0.05 g/m³) and thick (0.5 g/m³) fog.

thick fog (0.5 g/m³ humidity). It can be seen that fog is essentially invisible to microwave communications, and very low attenuation is experienced in the commercial millimeter-wave bands, up to around 100 GHz. However at higher bands, significant fog attenuation can result—up to almost 20 dB/km for thick fog at 1,000 GHz. Studies have shown that at optical frequencies, heavy fog with a visibility of 75m can exhibit 225 dB/km attenuation [8], effectively limiting communications of free-space optical transmission systems to just very short distances.

For ground to air applications, cloud attenuation needs to be considered. However cloud attenuation need only be considered over the segment of the path that may be prone to cloud cover (typically the lowest 10 km, above which cloud cover seldom occurs). The above analysis holds for both fog and clouds, although for cloud attenuations, the curves corresponding to 0°C should be used.

2.6.2 Snow and Ice

The effects of snow and ice on atmospheric attenuation at high frequencies have not been as well studied as those of rain and fog. Snowfall rates (in terms of water content) are generally less than rainfall rates, suggesting electromagnetic propagation is affected less by snowfall than by rain. However the size of snow and ice crystals is comparable to millimeter wavelengths, meaning that these forms of precipitation can cause atmospheric attenuation.

Reference [9] provides an overview of several experimental investigations and models based on snow crystals as hexagons. An experimental equation for calculation of the specific attenuation due to snow is given by:

$$\gamma_S = \frac{0.00349 R^{1.6}}{\lambda^4} + \frac{0.00224 R}{\lambda} \tag{2.16}$$

where R is the snowfall rate of melted water content measured in mm/hr and λ is the wavelength of interest. This yields values in the range of 0.2 to 1.0 dB/km, showing the minimal effect of flying snow on microwave and millimeter-wave transmissions.

Modeling attenuation due to ice can be more complicated, since the dielectric properties of ice are very different from those of liquid water. The dielectric losses of ice have a minimum near 1 GHz, but fortunately ice is a very nearly a lossless medium over a large frequency range. Therefore attenuation due to pure ice particles can be neglected in most situations. Special situations occur, however, when ice particles start to melt, since a very thin skin of liquid water surrounds the crystal, which can be sufficient to cause absorption.

2.6.3 Sand, Dust, and Other Small Particles

Sand and dust particles, as well as other small airborne particles (e.g., pollution), are much smaller than the water particles considered earlier, and much less than 0.1 mm in size. These particles are sufficiently small compared to microwave and millimeter-wave wavelengths that Rayleigh scattering should be considered. Even then, scattering would not be expected until up into the terahertz frequency bands. Since this is above all commercial wireless bands, and therefore not a consideration for commercial radio links, little study has been done in this area. However, it should be noted that commercially available free-space optical links do need to account for attenuation due to small particle effect. Severe outages can occur in sand and dust storms, similar to fog.

2.7 Attenuation Due to Blockage of the Transmission Path

High-frequency systems generally are line of sight, meaning that the radio cannot tolerate any physical blockage in the path between the transmitter and receiver. However high-frequency signals can, to some extent, penetrate foliage, although with attenuation depending on the depth of the obstruction. Furthermore, wireless signals can be diffracted off solid obstacles close, but not necessarily in line with the transmitter-receiver path, causing additional deterioration of the signal.

2.7.1 Foliage Losses

Losses due to foliage can be significant at high frequencies. Reference [10] outlines an empirical relationship, applicable for frequencies in the range of 200 MHz to 95 GHz and cases where the foliage depth is less than 400m, whereby the foliage loss L, measured in dB is given by:

$$L = 0.2 f^{0.3} R^{0.6} \qquad (2.17)$$

where f is frequency measured in megahertz and R is the depth of foliage measured in meters. As an example, for a penetration of 10m (equivalent to a large tree or two close together) a 40-GHz signal would have an attenuation of about 19 dB.

2.7.2 Diffraction Losses and Fresnel Zones

For lower-frequency transmissions, where paths tend to be longer and antennas have wider beamwidths, diffraction losses from obstacles such as buildings, towers, and even the curvature of the Earth need to be considered. For higher frequency transmissions, where paths tend to be shorter and high antenna gains keep transmissions highly directional and focused, diffraction fading is not usually an issue. It is nevertheless good practice to keep potential blocking obstacles a distance away from the line-of-sight path. The amount of clearance required by such obstacles is determined by Fresnel theory and stated in terms of Fresnel zones.

Reference [11] provides a detailed description of the theory of diffraction losses, Fresnel zones, and how they are applied to microwave transmissions. Fresnel theory, applied to wireless transmissions, defines a series of ellipsoids where the transmit and receiver antennas are at the poles of the ellipses. F_1 is the radius of the first Fresnel ellipsoid, measured in meters, and is given by:

$$F_1 = 17.3 \sqrt{\frac{d_1 d_2}{fd}} \qquad (2.18)$$

where f is the frequency measured in GHz, d is the path length measured in kilometers, and d_1 and d_2 are the distances from each antenna to the path obstruction (also measured in kilometers). Diffraction theory indicates that the direct path between the transmitter and the receiver needs a clearance of at least 60% of the radius of the first Fresnel zone to achieve free-space propagation conditions.

For a millimeter-wave transmission over a 1-mile link, the first Fresnel zone is typically only a meter or two at the midpoint of the link. Theoretically, since only 60% clearance of the first Fresnel zone is required for diffraction-free transmissions, this is only 1m or less. Clearly, obstacles can be situated very close to high-frequency wireless systems without causing diffraction losses. For this reason, high frequency systems can be used effectively in city environments to thread high data rate wireless transmissions between buildings and through obstacles.

2.8 Summary

Electromagnetic radio waves propagating through the Earth's atmosphere will encounter atmospheric attenuation, which will hinder transmission, causing outages or deterioration in wireless transmissions.

This chapter details the most significant atmospheric attenuation effects that need to be understood for high-frequency wireless transmission. These include free-space loss, attenuation due to atmospheric gases, attenuation due to precipitation, and attenuation due to other particles in the atmosphere such as fog, snow, ice, dust, and sand.

At frequencies above about 10 GHz, propagation through the atmosphere is greatly influenced by the effects of both molecular resonances and precipitation.

Empirical models have been presented for calculating the attenuation due to oxygen and water vapor. There are several resonance peaks due to water vapor and oxygen absorption that limit transmissions to short distances. There are also a number of atmospheric windows where atmospheric attenuation is sufficiently low that long distance transmissions can occur.

Losses from rain fades can be very significant for higher-frequency transmissions. Models for calculating atmospheric losses due to various form of precipitation including rain, snow, and fog have been provided. Losses through vegetation can also be significant, although diffraction losses for obstructions close to the path are not generally a problem for higher-frequency wireless transmissions.

References

[1] Thomas, E.G., and A.J. Meadows, *Maxwell's Equations and their Applications*, Bristol, UK: Adam Hilger, 1985.

[2] Westwater, E.R., S. Crewell, and C. Mätzler, "Surface-Based Microwave and Millimeter Wave Radiometric Remote Sensing of the Troposphere: A Tutorial," *IEEE Geoscience and Remote Sensing Society Newsletter*, 2005, pp. 16–33.

[3] ITU-R P.676-6, "Attenuation by Atmospheric Gases," 2005.

[4] Wells, J.A., "Faster Than Fiber: The Future of Multi-Gb/s Wireless," *IEEE Microwave Magazine*, Vol. 10, No. 3, 2009, pp. 104–112.

[5] ITU-R P.835-4, "Reference Standard Atmospheres," 2005.

[6] ITU-R P.838-3, "Specific Attenuation Model for Rain for Use in Prediction Methods," 2005.

[7] ITU-R P.840-3, "Attenuation Due to Clouds and Fog," 1999.

[8] Bloom, S., "The Physics of Free-Space Optics," *AirFiber White Paper*, 2002.

[9] Foessel, A., S. Chheda, and D. Apostolopoulos, "Short-Range Millimeter-Wave Radar Perception in a Polar Environment," *Proc. Field and Service Robotics Conference*, 1999.

[10] "Millimeter Wave Propagation: Spectrum Management Implications," Bulletin No. 70, FCC Office of Engineering and Technology, 1997. Reprinted as Marcus, M., and B. Pattan, "Millimeter Wave Propagation: Spectrum Management Implications," *IEEE Microwave Magazine*, Vol. 6, No. 2, 2005, pp. 54–62.

[11] Manning, T., *Microwave Radio Transmission Design Guide, Second Edition*, Norwood, MA: Artech House, 2009.

3

High Data Rate Wireless Systems

3.1 Introduction

A very simplistic representation of a generic wireless system is shown in Figure 3.1. Here unit A and unit B are connected together via a wireless signal propagating through free space. The over-the-air wireless signal will conform to a particular wireless protocol, enabling the two units to transmit and receive information and intelligently understand and interpret the communication accordingly. This wireless protocol may be proprietary, meaning the communications protocol used is not a recognized standard. In such cases, units A and B are likely to be manufactured and sold as a pair from a single manufacturer or vendor, and are not interoperable with other vendors' equipment. An example of this might be a fixed point-to-point (PTP) radio system for connecting together two buildings or two devices, whereby the system consists of two paired units with a dedicated high-speed communication channel between the two. Most high data rate communications systems are PTP. A characteristic of high-speed PTP communication systems is that the wireless path needs to be line-of-sight.

Alternatively the wireless protocol may be standards-based. Here unit A and unit B are not necessarily manufactured or sold by the same vendor, but the two units are sold to operate together. Examples of such wireless protocols are the various cellular standards (the wireless protocol could be GSM, CDMA, LTE, or any one of many other standards, and unit A and unit B might be a cell phone and a base station, respectively), or wireless local area/wide area networks

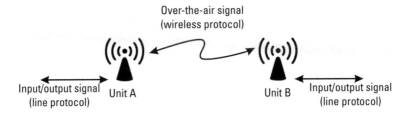

Figure 3.1 Generic diagram of a simplified wireless link.

(WLAN/WAN) standards [the wireless protocol may be WiFi (IEEE 802.11) or WiMAX (IEEE 802.16), and unit A might be a laptop or customer premise equipment (CPE) and unit B might be an appropriate access point or base station]. As such, most standards-based communication systems are point-to-multipoint (PTMP), where communication is from a single device to two or more other devices. Since the over-the-air capacity is shared between many individual paths, true high data rates cannot be achieved with PTMP systems.

This chapter details the various architectures, subassemblies, and protocols that make up a high data rate wireless system. Device architectures are broken down into various subassemblies, and it is shown how these are practically configured. Detailed descriptions of the various line protocols that connect to the wireless devices are given. Finally, the individual functional blocks that make up the radio architecture are presented and discussed, showing the purpose and key characteristics of each functional subassembly and key device within the overall wireless system.

3.2 High Data Rate Wireless System Architectures

High data rate radios come in several forms to satisfy many different wireless applications. It is useful to classify wireless systems by their physical structure or architecture.

3.2.1 Indoor and Outdoor Architectures

Figure 3.2 shows a conceptual block diagram of a wireless system, showing that each radio unit consists of essentially three parts: the signal processing unit (SPU), the radio frequency unit (RFU), and the antenna. The signal processing unit (SPU) inputs the high-speed digital line signal, manipulates the signal, adding various signal processing coding and modulation functions, and outputs either a baseband or IF (intermediate frequency) signal for the radio frequency unit (RFU). The RFU takes the output signal from the SPU and converts this to a higher carrier frequency, with a minimum of added noise and other electronic impurities. The signal is amplified and transmitted via the radio antenna. Since

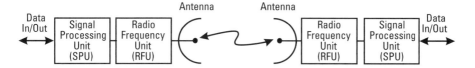

Figure 3.2   Block diagram of two simplified wireless units forming a wireless link.

each unit performs both a transmit and a receive function, a reverse receive path also occurs. Here the antenna receives the signal, and the RFU detects the signal and downconverts to the SPU with a minimum of added noise and electrical impurities. The SPU then extracts the communication signal information of interest and outputs this data to the rest of the network.

3.2.1.1   All-Indoor Radios

Some high data rate radios are used exclusively for indoor use. Usually the goals for such equipment are cost, size, weight, and ease of installation, commissioning, and use. An example of such an indoor residential wireless system is a 60-GHz wireless unit used for delivering high-definition audiovisual content around a living room, as shown in Figure 3.3. In industrial infrastructure applications, some wireless units are also used indoors. For terrestrial trunked microwave links, all-indoor microwave units are commonly used. Here the RFU and SPU reside indoors, with the antenna mounted outside, usually high on a tower,

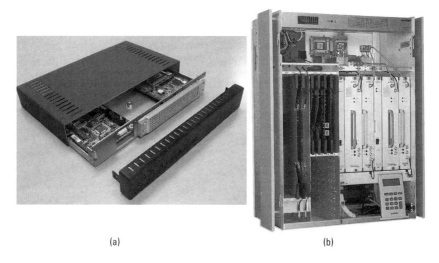

(a)   (b)

Figure 3.3   Photographs of two all-indoor high data rate wireless systems. (a) 60-GHz system supporting up to 4 Gbps of audiovisual data for in-home WirelessHD transmission applications. (*Source:* SiBEAM, Inc., 2010. Reprinted with permission.) (b) All-indoor microwave system supporting protected OC-3 data streams for long-distance trucking applications. (*Source:* Aviat Networks, 2010. Reprinted with permission.)

connected via a low loss waveguide pipe. A picture of such a unit is also shown in Figure 3.3. Key design goals are usually reliability and performance. Typically, such units operate in the lower microwave bands (for example, 6 and 11 GHz) and are used for high capacity data transmission for long-distance backbone trunking applications.

3.2.1.2   All-Outdoor Radios

Other applications demand all-outdoor unit operation. Here the wireless system is mounted on a rooftop or attached to a tower next to the antenna. All-outdoor radios are easy to install and use no space in equipment rooms. However, all-outdoor radios require long cable runs as power and data interfaces have to be run from the peripheral equipment up to the rooftop or up the tower for interconnection to the radio. Many all-outdoor radios incorporate an integrated antenna. Since an all-outdoor radio resides in the open air, it is exposed to much wider temperature variations than an indoor radio, which is usually mounted in a temperature- and humidity-controlled environment (inside a house or a telecom hut). The radio is also exposed to potentially harsh weather elements (rain, snow, icing, salt spray, and lightning), which can easily damage the radio or degrade its performance. Key design goals are therefore usually focused on reliability. A typical all-outdoor radio is shown in Figure 3.4.

3.2.1.3   Split Mount Radios

Split mount radios are configured with part of the system outdoors and part of the system indoors. Usually the SPU is mounted indoors, and renamed an indoor unit (IDU). The RFU is mounted outdoors and renamed an outdoor unit (ODU). The two are usually connected together via a coaxial cable, which is used to pass signals between the IDU and ODU. This cable carries the transmit

Figure 3.4 Picture of all-outdoor high data rate wireless system. Shown is a 1-Gbps 71–86-GHz point-to-point radio mounted to a 1-ft antenna. (*Source:* E-Band Communications Corporation, 2010. Reprinted with permission.)

and receive IF signals between the IDU and ODU, provides the IDU to ODU communication telemetry and control signals, and passes the DC power to the ODU. Since the ODU is attached to the antenna, feeder losses are minimized and output power is maximized. Since the IDU is mounted indoors, the data interface and management ports are easily accessible and configurable, adding to the flexibility of the system. Split mount systems are therefore relatively easy to install and maintain and quick to configure and commission. The split mount format is the most widely used equipment configuration in modern-day microwave equipment. A typical split mount system is shown in Figure 3.5.

3.2.2 Diplexing Architectures

3.2.2.1 Simplex and Duplex

In a general communication system, two connected devices communicate with one another. If the communication is unidirectional (for example, in a broadcast application where one device is required to only transmit and another only receives), the operation is referred to as a simplex communication system. If the

Figure 3.5 Picture of split mount wireless system. Shown is a high capacity microwave split mount radio consisting of an IDU, an ODU, and an antenna. Three ODUs connected to 1-ft (30-cm) antennas are shown to illustrate the ODU form factor. (*Source:* Aviat Networks, 2010. Reprinted with permission.)

two devices communicate together in a bidirectional path, the system is referred to as a duplex communication system.

There are two types of duplex schemes. A half-duplex system provides for communication in both directions, but only one direction at a time. Typically, once a device begins receiving a signal, it must wait for the transmitter to stop transmitting before replying. A full-duplex system permits communication simultaneously in both directions. To achieve this, the full-duplex system requires two physically separate channels for system operation. Full-duplex is far more efficient than half-duplex, since the full data capacity of the channel is available in both directions as the send and receive functions are separate. There are no timing limitations whereby there is downtime waiting to determine if transmissions have ended, and there is no need for retransmissions as there are no collisions on the separate channels.

All the high data rate wireless systems considered in this book operate as duplex systems. In the wireless communications world, these are named time division duplex (TDD) and frequency division duplex (FDD). Conceptual operation of both TDD and FDD is shown in Figure 3.6.

3.2.2.2 Time Division Duplex (TDD)

TDD is the application of time division multiplexing to separate outward and return signals on a single frequency channel. TDD is especially advantageous where there is asymmetry on the data and the traffic can be dynamically al-

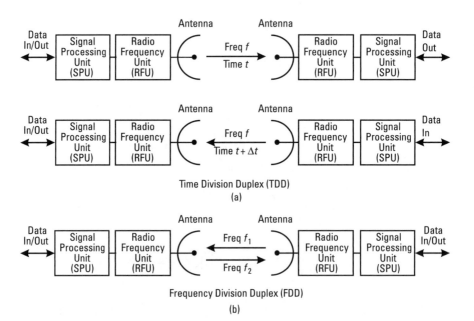

Figure 3.6 Conceptual operation of (a) TDD and (b) FDD configurations.

located to provide more capacity on the uplink or downlink, depending on the traffic flow. TDD is more common in lower capacity applications, especially in the unlicensed frequency bands, where limited frequency channels and regulatory control are implemented. For a number of reasons, TDD is also not usually preferred for high capacity transmissions, as it is inherently inefficient in managing data capacity. First, the full throughput capacity of the link has to be shared between transmissions in both directions. Thus, for a symmetrical link, only 50% of the traffic capacity occurs in each transmission direction. Second, this 50% value can never be achieved in practice. TDD systems need to be implemented with guard times between transmissions, to allow switchover from transmit to receive. This downtime decreases effective network throughput. In addition, TDD has a number of implementation complexities. Advanced signal processing techniques with precise timing are required to manage and synchronize high-speed switching, which add cost and latency penalties. Very high-speed switches are required to control rapidly changing traffic flows.

3.2.2.3 Frequency Division Duplex (FDD)

FDD is widely used in the licensed microwave and millimeter-wave bands, where regulators deliberately allocate pairs of frequency channels to use together. As such, the radio's transmitter and receiver operate at different carrier frequencies, separated by the TR spacing (transmit-receive spacing). FDD is every efficient in throughput, as no consideration of switching or directional timing is required. Thus, the full radio throughput is available for carrying meaningful data traffic. FDD also makes radio planning much easier and more efficient, since devices transmit and receive in different subbands and therefore have minimal risk of interfering with one another. All high capacity infrastructure links operate in an FDD configuration.

Although usually symmetrical, FDD can be operated asymmetrically with a primary, higher data rate path and a secondary, lower data rate return. FDD is hard to dynamically balance like TDD, but can be used effectively for systems with a low data rate "back channel" for monitoring and control of remote equipment.

It is conventional to quote an FDD system's throughput as its data carrying capacity for just one of the two carrier frequency channels. For example, a 1-Gbps FDD link is by convention carrying two 1-Gbps data streams, one transmitting from device A to device B and the other simultaneously transmitting from device B to device A. Thus, the over-the-air data rate is actually 2 Gbps, all of which is available for communication traffic. A 1-Gbps TDD link, however, has by convention only 1 Gbps of over-the-air data to allocate for transmission between device A and device B. This could be 1 Gbps solely in just one direction, 500 Mbps in both directions, or some other combination. Because of the control data that needs to be sent over the air to manage the switching data connections and the guard times required to ensure proper timing, a typical 1-Gbps TDD system might only be 75% to 85% efficient, meaning that only 750 to 850 Mbps of the 1 Gbps is available for communication traffic. However, a 1-Gbps FDD system actually has a full 2 Gbps of capacity available for communication traffic.

3.3   High Data Rate Line Protocols

To connect a high-speed wireless link with a wider network, high-speed input/output signals are necessary. These signals will conform to a high-speed, standards-based line protocol to enable interconnectivity with a variety of external peripheral equipment. In the wireless world, there are many high-speed line protocols, which are considered in this section. The transmission of these high-speed signals is usually via a high-speed wired medium such as fiber optic cable, coaxial cable, or unshielded twisted pair (UTP) category 5 or 6 (CAT-5, CAT-6) cable.

3.3.1   PDH

Plesiochronous digital hierarchy (PDH) is a transmission technology widely used in both wired and wireless systems. The term *plesiochronous* is derived from the Greek *plesio*, meaning near, and *chronos*, which means time, and refers to the fact that PDH networks run in a state where different parts of the network are nearly, but not quite perfectly, synchronized. PDH systems differ slightly in America and Europe, but the basic principles are the same. A base channel of 64 kbps is defined (historically the channel size needed to accommodate one 8-kHz voice signal using an 8-bit analog-to-digital converter). In the United States, 24 of these base channels are aggregated together, along with some small signaling overhead, to form a T1 (or DS1) of data rate 1.544 Mbps. In Europe, 32 of the base 64-kbps channels are aggregated to form a 2.048 Mbps E1 circuit, two channels of which are usually used for signaling. Although originally designed for carrying voice traffic, the PDH signal can be used for nonspeech purposes such as data transmission.

For higher data rate applications, multiple T1 or E1 circuits are combined (multiplexed) together, usually in groups of four, to allow higher data rates to be accommodated. This hierarchy is shown in Table 3.1. The terms T1 and DS1 are often used interchangeably, although, strictly speaking, a DS1 is the data carried on a T1 circuit. The same is similarly for other T/DS rates.

3.3.2   SDH/SONET

For high capacity wireless data transmission, synchronous digital hierarchy (SDH) or synchronous optical networking (SONET) protocols are more com-

Table 3.1
PDH Levels, Data Rates, and Designations for Both
North American (T-Rate) and European (E-Rate) Systems

Level	North American PDH Designations and Data Rates		European PDH Designations and Data Rates	
Zero (base channel)	DS0	64 kbps		64 kbps
First level	DS1 or T1 (equal to 24 base channels)	1.544 Mbps	E1 (equal to 32 base channels)	2.048 Mbps
Second level (not widely used)	DS2 or T2	6.312 Mbps	E2	8.448 Mbps
Third level	DS3 or T3 (equal to $28 \times$ T1)	44.736 Mbps	E3 (equal to 16 $\times$ E1)	34.368 Mbps
Fourth level (not widely used)	DS4	274.176 Mbps	E4	139.264 Mbps
Fifth level (not widely used)	DS5	400.352 Mbps	E5	565.148 Mbps

mon than PDH. SDH is a European designation, and SONET is the North American designation. Although named differently, both SDH and SONET are very similar in structure, and both provide a path to much higher data rates. Both are widely used in telecommunications equipment.

SDH/SONET was developed in the late 1980s to be a unified standard that allows better interoperability between different vendors equipment, provides a path to higher data rates, and removes some of the limitations of lower capacity PDH protocols. SDH/SONET works by aggregating either the E-rate and T-rate PDH frames and providing synchronization and overhead frames that allow for better network manageability and more efficient demultiplexing. Table 3.2 shows the hierarchy of available SONET/SDH data rates.

Work is underway to standardize the next logical rate of 160 Gbps (OC-3072/STM-1024). All the other data rates are widely implemented in fiber optic systems. The lower SONET/SDH data rates to 155 Mbps have been widely implemented in radio systems, particularly for telecommunications applications. Recent technical developments have enabled 622 Mbps (OC-12/STM-4) and higher systems to be realized wirelessly.

3.3.3   Ethernet

Ethernet is a transmission protocol widely used in wired communications, which has recently been adopted by the wireless community. Ethernet was developed in the early 1970s, and has evolved into the most widely implemented protocol today. Various iterations include Fast Ethernet, which increased Ethernet data speeds from 10 Mbps to 100 Mbps, and Gigabit Ethernet (GbE), which further increased speeds to 1 Gbps.

Table 3.2
SDH/SONET Data Rates and Designations

North American SONET Designation	European SDH Designation	Data Rate
OC-1	STM-0	51.84 Mbps
OC-3	STM-1	155.52 Mbps
OC-12	STM-4	622.08 Mbps
OC-24	—	1.244 Gbps
OC-48	STM-16	2.488 Gbps
OC-192	STM-64	9.953 Gbps
OC-768	STM-256	39.813 Gbps

The base standard for gigabit Ethernet is IEEE 802.3z, released in 1998. This is commonly referred to as 1000BASE-X, where X refers to various designations associated with the physical layer/transmission medium (for example, fiber or cable).

For telecommunications networks, where fiber optics is often the transmission technology of choice, 1000BASE-SX and 1000BASE-LX are relevant. These are defined in IEEE 802.3z for transmission over multimode and single-mode fibers, respectively. This standard employs 8B/10B encoding (every 10 bits sent carry 8 bits of data), which increases the line rate from 1.0 Gbps to 1.25 Gbps to ensure a DC-balanced signal, but with sufficient state changes to allow a recoverable clock.

In the commercial business or enterprise environment, 1000BASE-T is more relevant. This is defined in IEEE 802.3ab for transmission over unshielded twisted pair (UTP) category 5 or 6 (CAT-5, CAT-6) cabling. This uses a different encoding scheme to the fiber standards, in order to keep the symbol rate as low as possible for the short-distance, limited data capacity cables. 802.3ab gigabit Ethernet has become popular as enterprises can utilize their existing lower capacity Ethernet copper cabling infrastructure. In 2000, Apple's Power Mac G4 and PowerBook G4 were the first mass produced personal computers featuring the 1000BASE-T connection. Gigabit Ethernet is now a common built-in feature in computers, with 1000BASE-T network cards included in almost all desktop and server computer systems. A summary of the Gigabit Ethernet modes is given in Table 3.3.

Faster 10-Gbps Ethernet (10GbE) standards are now available, and have been consolidated into the IEEE 802.3 set of standards. 10GbE is now replacing

Table 3.3
Gigabit Ethernet Modes and Their Characteristics

IEEE Designation	Description	Cabling	Distance
1000Base-LX (IEEE 802.3z)	Long haul fiber	9 micron single-mode fiber	Up to 5 km
1000Base-SX (IEEE 802.3z)	Short haul fiber	62.5 micron multimode fiber, or 50 micron multimode fiber	Up to 275m or 550m, respectively
1000Base-T (IEEE 802.3ab)	Long haul copper	UTP CAT5 and CAT6	100m
1000Base-CX (IEEE 802.3z)	Short haul copper	Shielded copper	25m

1GbE in the fiber backbone network, is migrating to high-end server systems, and is being seen as a backplane of high performance equipment.

Work is underway within standards bodies on developing 40 Gbps and 100 Gbps Ethernet standards.

3.3.4 Other High Data Rate Line Protocols

3.3.4.1 InfiniBand

InfiniBand is a gigabit-speed standard used primarily in high-performance computing, connecting together processors with high-speed peripherals such as disks. As such, it is a short-distance wired technology. InfiniBand's base data rate is 2.0 Gbps. Like Gigabit Ethernet, links use 8B/10B encoding, which adds 25% coding overhead, making the line rate 2.5 Gbps. InfiniBand supports double and quad data speeds for 5 Gbps and 10 Gbps, respectively. InfiniBand is typically used for supercomputer interconnections and interswitch connections.

3.3.4.2 Fiber Channel

Fiber Channel is a North American accredited gigabit-speed network technology primarily used for storage area networks (SAN). Despite its name, Fiber Channel can run on both twisted pair copper wire and fiber-optic cables. Interoperable Fiber Channel standards are defined at multiples of 1 Gbps, including 1, 2, 4, and 8 Gbps. Higher data rate standards at 10 Gbps and 20 Gbps are defined, but are not backwards compatible with the lower data rates.

3.3.4.3 SMPTE Standards

The Society of Motion Picture and Television Engineers (SMPTE) has over 400 standards and guidelines for television, motion pictures, digital cinema, and other applications. These include a set of Serial Digital Interface (SDI) standards for the transmission of gigabit-speed, uncompressed, unencrypted, broadcast-grade digital video signals over coaxial cable, usually within television and film studios.

A list of standards and data rates is given in Table 3.4. At the gigabit speeds, the high-definition serial digital interface (HD-SDI) is standardized in SMPTE 292M and provides a nominal data rate of 1.485 Gbps. The emerging dual link HD-SDI, consisting essentially of a pair of SMPTE 292M links providing 2.970 Gbps transmission, is standardized in SMPTE 372M. This has applications in digital cinemas or 1080p high definition TVs that require greater fidelity and resolution than standard HDTV. A more recent interface, 3G-SDI, consisting of a single 2.970 Gbit/s serial link, is standardized in SMPTE 424M and will likely replace the dual link HD-SDI.

3.3.4.4   OBSAI and CPRI

Two new gigabit data rate standards emerging in the telecom field are Open Base Station Architecture Initiative (OBSAI) and Common Public Radio Interface (CPRI). The two are competing standards, both developed with the same goal: to standardize cellular base station architecture in a way that substantially reduces development efforts and costs. Both standards define a high data rate interface to go between a modularized baseband SPU modem module and the RFU module. The OBSAI specification allows for line rates of $n \times 768$ Mbps, where $n = 1, 2$, and 4 (768 Mbps, 1.536 Gbps and 3.072 Gbps). CPRI similarly defines high data rate interfaces, but at a different line rate: $m \times 614.4$ Mbps, where $m = 1, 2, 4, 5, 8$, and 10 (614.4 Mbps, 1.2288 Gbps, 2.4576 Gbps, 3.072 Gbps, 4.9152 Gbps, and 6.144 Gbps). The standards envision this data being sent via an optical fiber link. However, with the emergence of Distributed Antenna Systems (DAS) and remote radio heads, there is a need to locate the RF unit away from the core of the base station. The CPRI and OBSAI standards will allow for this, with an opportunity for wireless to transmit the high data rates rather than existing fiber technologies.

Table 3.4
SMPTE Standards, Associated Data Rates, and
Example Video Formats Supported by the Standard

Standard	Name	Bit Rates	Example Video Formats
SMPTE 259M	SD-SDI	270 and 360 Mbps	480i, 576i
SMPTE 344M	—	540 Mbps	480p, 576p
SMPTE 292M	HD-SDI	1.485 Gbps	720p, 1080i
SMPTE 372M	Dual Link HD-SDI	2.970 Gbps	1080p
SMPTE 424M	3G-SDI	2.970 Gbps	1080p

3.4 High Data Rate Wireless Building Blocks

High data rate radio systems come in a variety of different architectures for different applications. Figure 3.7 shows a generic radio architecture for a high data rate wireless equipment.

The SPU serves as the high-speed data signal input and output interfaces and performs the functions of management and control access, multiplexing, modulation, and demodulation. The RFU provides the upconversion and downconversion of the signal, amplification of the transmit signal, and low noise reception of the receive signal. This is all performed with a minimum of added spectral impurities and noise that would impair the signals. The antenna is used to covert the electronic signal to an electromagnetic radio wave for both transmission and reception.

Each of the SPU and RFU subassemblies is detailed in this section to demonstrate their functionality and key characteristics for high data rate wireless transmission.

3.4.1 Power Supply

The power supply is an often overlooked, but critically important part of a wireless system. The power supply has to provide a convenient power input that is converted to generate all the internal voltages necessary to power and operate the radio. Depending on the application, the power supply can be an AC or DC input. The power supply has to be high efficiency to avoid too much power being lost and be highly reliable.

One power supply performance requirement often addressed late in a design is electromagnetic capability (EMC). This is the ability for the wireless

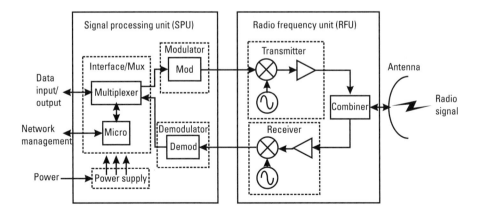

Figure 3.7 Generic high data rate radio block diagram.

system to operate with sufficiently low emissions (both radiated and conducted) so as to not interfere with other nearby equipment. Outside the United States, stress is also placed on meeting immunity specifications, ensuring that equipment can operate in the presence of unwanted, potentially interfering signals. Since the power supply is the mechanism by which the system interfaces (power-wise) to the rest of the network, the power supply needs to be designed to tolerate poor quality input power with high levels of noise and/or transient spikes and also has to avoid feeding back noise or interference from the wireless system back to the connecting network.

3.4.2   Data Interface/Multiplexer

A block diagram of the interface/multiplexer section is shown in Figure 3.8.

The high data rate input/output customer tributaries can be of a wide variety of formats, depending on the application. Figure 3.8 shows a hybrid radio with both time division multiplex (TDM) interfaces such as PDH or SDH and Ethernet interfaces. TDM signals are buffered via line interface units (LIUs) that provide the tributary line drivers and rate encoders, any necessary clocking and impendence matching, plus additional functions such as low frequency jitter and wander attenuation and local loopback functions. For any Ethernet interfaces, a packet processor such as a switch or router is usually placed in line with the tributary interface to provide higher-level networking and routing capabilities. These LIUs and switches also help ensure that the interfaces meet regulatory specifications and interoperability requirements.

The multiplexer combines all these customer tributaries together into a single digital bit stream, known as a frame. The format of this frame structure can be either proprietary or nonproprietary, depending on the data traffic being sent and the application in which the system is being used. If the system is Ethernet-only, the Ethernet frame structure may be preserved. If the system is TDM

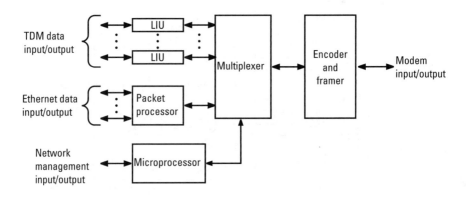

Figure 3.8 Block diagram of interface/multiplexer architecture.

only, a TDM-type frame structure might be used. If traffic is a combination of the two, or any other type of tributary traffic, then the Ethernet traffic might be encapsulated in a TDM frame, or the TDM traffic might be encapsulated in an Ethernet frame. Formal standards exist to describe such encapsulation techniques. For example, the pseudowire standard PWE3 (pseudowire emulation edge to edge), which was set up and championed by the Internet Engineering Task Force (IETF), defines such an encapsulation technique. Other standardization forums, including the ITU, are also active in producing standards and implementation agreements for similar encapsulation standards.

Once the multiplexer has combined all the customer tributary data in a suitable format, various other forms of data traffic are added to the multiplexed customer tributary signals. For example, the frame may include additional communication information from auxiliary data channels (such as a wayside serial channel that allows external controllers to send messages across the wireless link) and additional network management data that enables the microprocessor on one side of the radio link to communicate with the microprocessor on the other end of the radio link. All this traffic—the customer data and the network and management overhead—is multiplexed together into a single stream of data.

At this stage, various additional techniques are employed to add robustness to the data stream and improve its immunity to interference and other wireless channel impairments. Forward error correction (FEC) is one such technique widely used to improve the robustness of wireless transmissions and preserve the integrity of the transmitted data. FEC deliberately adds redundant bits of information to the data transmission, that the receiver is able to detect and correct (within some limits) without the need to ask the transmitter for additional information. Thus, FEC does not require a return "back channel" and retransmission of data can be avoided.

In its simplest form, FEC adds additional bits of information, usually in the form of parity bits, which describe the rest of the frame structure. For example, parity bits can be added prior to transmission that describe whether the frame has even or odd numbers of ones and/or zeros. At the receiver, the system can interrogate these parity bits and determine whether the relationship is still preserved. If it is, the system can pass the data, knowing with a certain level of confidence that the data received is the same as the data sent. If the relationship does not hold, algorithms can be applied that are able to isolate and correct many of the errors. The strength of the FEC depends strongly on the implementation. Popular FEC techniques are block codes, such as Reed-Solomon, which is implemented in CDs, DVDs, and hard disk drives and widely used in radio systems, and more powerful convolution codes such as the Viterbi code. Recently, low-density parity check (LDPC) coding has experienced a resurgence for wireless systems. Although FEC adds robustness to the data transmission, it does add significant overhead, processing complexity, and latency to the system. However, the strong improvements in system performance usually outweigh these disadvantages, and FEC is widely employed in digital wireless systems.

A typical radio frame structure is shown in Figure 3.9. This frame is representative of the data transmitted over the air by the wireless link. It can be seen that the data contained here is clearly greater than the original customer data. Thus, the aggregate over-the-air data rate of a wireless system is always larger than the actual data rate available for data use by the system user.

Not shown in Figure 3.9 is interleaving. Once the frame structure is defined, it is often deliberately muddled to add further robustness to the transmission. For example, if a wireless link is subject to a sharp burst of interference such as impulse noise, it can corrupt a continuous group of bits such that the FEC can no longer correct the errors. By deliberately spacing adjacent bits of information out across the frame in a predetermined fashion, the likelihood of large blocks of continuous data being corrupted is reduced, making the FEC more effective. Similar to FEC, interleaving adds robustness to the data transmission, but at the penalty of processing complexity and added latency to the system.

A final technique commonly seen on PTMP wireless links is encryption. This is not widely used on high capacity communication links since the radios are usually engineered to provide maximize transmission speed between two particular points. In addition, the risk of nonmalicious interception is low, since high capacity links are usually PTP and configured with high directional gain antennas.

It should be noted that the interface, multiplexer, encoder, and framer functions are reciprocal, in that the functions described above for transmitting high-speed customer tributary data are used in reverse at the other end of the wireless link to receive and regenerate the customer tributaries. Practically, the multiplexer, encoder, and framer functions are usually implemented in a structured or programmable signal processing device [for example, a field programmable gate array (FPGA) or application specific integrated circuit (ASIC)] programmed or specifically designed to perform the various signal processing functions. Often the modem function, described next, is also implemented digitally in the same FPGA or ASIC device.

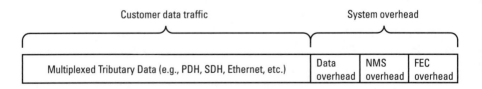

Figure 3.9 Example of typical radio frame.

3.4.3 Modulator

Thus far, the signals considered are digital baseband signals. For the baseband signal to be transmitted across a wireless link, it needs to be converted into an analog signal suitable for transmission. This conversion is done by superimposing the digital baseband signal, containing the customer data and management overhead traffic, onto a higher frequency analog carrier signal that can be transmitted. The process of transferring digital information into an analog form and compressing it into a given bandwidth is known as modulation, as shown in Figure 3.10.

As shown earlier, an electromagnetic wave E can be described by the following equation, which shows the signal is a function of three variables: amplitude (E_0), frequency (f), and phase (ϕ):

$$E = E_0 \cos\left(2\pi f t + \phi\right) \tag{3.1}$$

Each of these three parameters can be varied (modulated) in a manner that enables the high frequency carrier to convey the digital information required to be communicated across the wireless link. Thus, the following classes of digital modulation are widely used in high capacity wireless links:

- *Amplitude shift keying (ASK):* Modulating the carrier signal's amplitude, but keeping the signal's frequency and phase constant;
- *Frequency shift keying (FSK):* Modulating the carrier signal's frequency, but keeping the signal's amplitude and phase constant;
- *Phase shift keying (PSK):* Modulating the carrier signal's phase, but keeping the signal's amplitude and frequency constant;
- *Quadrature amplitude modulation (QAM):* Modulating the carrier signal's amplitude and phase, but keeping the signal's frequency constant.

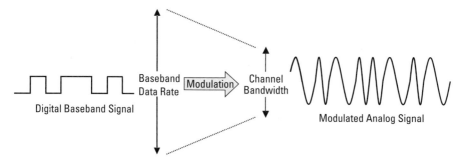

Figure 3.10 The modulation process: superimposing a digital signal onto an analog carrier and compressing to fit in a limited size channel bandwidth.

Each of these modulation schemes has advantages and disadvantages for high data rate wireless communications. Each is summarized next and in Figure 3.11.

3.4.3.1   ASK Modulation

ASK is the digital equivalent of analog amplitude modulation (AM). Here the amplitude of the signal is switched between two amplitude levels reflecting the transmission of a binary one or zero. In its simplest form, the presence of a transmitted signal can be used to convey a binary one, and the absence of a signal (effectively zero amplitude) can be used to represent a binary zero. In this case the modulation is called on-off keying (OOK). This is a popular technique for very simple transmission systems, as both the transmitter and receiver are relatively simple architectures and straight forward to realize. However, this modulation system is not spectrally efficient, meaning wide channel bandwidths are required

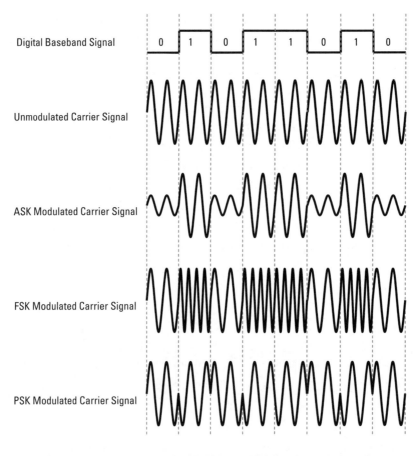

Figure 3.11 Example implementations of ASK, FSK, and PSK digital modulation schemes.

to transmit high data rates. It is also prone to noise interference. Therefore, OOK is widely used in optical communications where there are very wide bandwidths available and controlled levels of noise. It is also popular for lower cost wireless systems in the wider channel millimeter-wave bands.

3.4.3.2 FSK Modulation

FSK is the digital equivalent of analog frequency modulation (FM). Here the frequency of the carrier signal is switched between two values to reflect the transmission of a binary one or zero. FSK is not very spectrally efficient, but it can be used in a multilevel form whereby four or more frequency levels are used to convey more than 1 bit of information at a time (e.g., 4-FSK uses four discrete frequency steps to convey twice the information rate than conventional two-level FSK). The big advantage of FSK modulation is that it is a constant envelope modulation scheme with no information contained in the amplitude of the transmitted signal. Therefore, power amplifiers can be run into saturation, meaning that power can be generated relatively cheaply and easily. Also, since there is also no information in the phase of the signal, the noise and stability of frequency sources are not critical. Therefore, FSK is both cost-effective and robust and widely used in lower cost radio systems. Variations of FSK modulation are used in GSM cell phones. FSK radios are also widely used in lower capacity radio systems, but their limited spectral efficiencies and inability to scale makes them of limited use for high data rate wireless communications.

3.4.3.3 PSK and QPSK Modulation

PSK is the digital equivalent of analog phase modulation (PM), whereby the phase of the carrier signal is switched to reflect the transmission of digital information. The simplest form of PSK is binary phase shift keying (BPSK), whereby the phase of the signal is either unchanged (0° phase shift) or reversed (180° phase shift) to convey either a binary zero or one. Like FSK, PSK is a constant envelope modulation and can make efficient use of transmitter power. It is also easily scalable to higher levels, bringing increased spectral efficiency. For example, quadrature phase shift keying (QPSK), which is equivalent to 4-PSK, uses four discrete phase states to convey twice the information of regular BPSK. This can be best seen by viewing the PSK modulation on a phase or constellation diagram. Consider Figure 3.12. In Figure 3.12(a), the two phase changes associated with BPSK modulation can be seen, whereby the carrier signal changes 180° when representing a digital one or zero. In Figure 3.12(b), the four possible phase changes for QPSK can be seen, each 90° relative to each other, representing two binary bits of information per phase change. Thus, in contrast to a BPSK system that considers each individual binary bit to make a corresponding phase change decision, a QPSK system looks at pair of binary bits and then makes a finer phase change decision based on the information contained in this pair of bits.

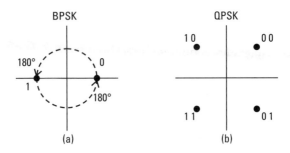

Figure 3.12 Constellation diagrams of (a) BPSK and (b) QPSK modulation.

Therefore, QPSK can convey two bits of information per phase change, making it twice as spectrally efficient as BPSK.

QPSK is widely used in communication systems, where relative simplicity, robustness, and spectral efficiency are required. QPSK can be implemented in various forms. For example, differential QPSK (DQPSK) requires each phase change to be relative to the current phase state, rather than absolute to a fixed constellation scheme each time. Offset QPSK (OQPSK) is a more complicated implementation that limits phase changes to no more than 90°, reducing large amplitude variations that occur in practice as the carrier signal swings between large phase changes. $\pi/4$-QPSK is a further scheme that employs two constellations offset by 45° ($\pi/4$ radians) to further limit phase transitions.

PSK modulation can be extended beyond DPSK and QPSK to encompass any number of phase changes. When viewed on a phase diagram, the constellation points are usually chosen with uniform angular spacing around a circle, giving a maximum phase-separation between adjacent points, enabling the receiver to easily distinguish between different phase changes and regenerate correct data. Also constellation points are spaced on a circle to yield similar magnitudes so that all phase changes will be transmitted with the same energy. Thus, PSK is a robust modulation scheme with good immunity to noise and corruption.

3.4.3.4   QAM Modulation

QAM is an extension to PSK modulation that brings a key benefit – spectral efficiency. For high data rate systems, it is usually desirable to fit as much data as possible into a finite-sized channel bandwidth. For this reason, QAM is widely employed in high data rate wireless systems.

QAM operates by modulating the amplitude of two PSK carriers 90° out of phase with each other to convey the required digital data signal. The two carriers are said to be in-phase (I) and quadrature (Q). Similar to QPSK, it is instructive to view QAM signals on a constellation diagram where the axes are the I and Q carriers, as shown in Figure 3.13. In fact, QPSK is really the simplest version of QAM, whereby the modulating signal is constant and just the phase

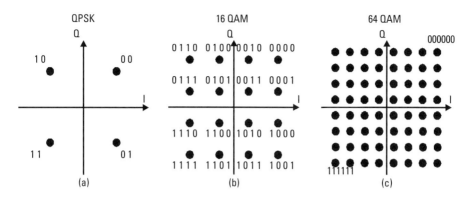

Figure 3.13 Comparison of (a) 4-QAM (QPSK), (b) 16-QAM, and (c) 64-QAM constellations.

is varied. For this reason, QSPK is often called 4 QAM. For QAM signals, the constellation points on the phase diagram are usually arranged in a square grid with equal vertical and horizontal spacing. Since the data in digital systems is binary, the number of points in the grid is a power of 2. The system is therefore called an M-ary system where $M = 2^n$ and n is the number of bits per symbol. n is also known as the modulation level or the modulation efficiency, measured in b/s/Hz. In the constellation IQ diagram, M is equal to the number of constellation points, and n is the number of bits represented per constellation point. For example, 64 QAM is a six-level modulation scheme with a theoretical efficiency of 6 b/s/Hz, since it allows the transmission of 6 bits per symbol ($64 = 2^6$). 64 QAM has 64 unique modulation states, and so is represented by an 8×8 square constellation on the IQ phase diagram.

It can be seen that by moving to higher-order modulations, it is possible to transmit more bits per digital symbol and hence gain higher modulation efficiency. However, as the modulation efficiency increases, the constellation points move come closer together and thus become harder to differentiate and decode and therefore more susceptible to noise and other corruptions. In addition, the receiver must correctly detect both phase and amplitude, adding considerable complexity to the demodulator design and raising the risk of bit errors. Nevertheless, in general, QAM achieves a greater distance between adjacent constellation points by distributing the points more evenly than other modulation schemes, meaning that overall QAM is more robust than similarly comparable alternatives.

It is convenient to use the term symbol rate when dealing with M-ary modulated systems. Since each phase and amplitude change can convey a different number of bits depending on the modulation complexity, the term symbol rate often has more relevance to system design than the bit rate. The symbol rate F_S is related to bit rate F_B by:

$$F_S = \frac{F_B}{\log_2 M} = \frac{F_B}{n} \tag{3.2}$$

As an example, consider a 256 QAM system transmitting a 350-Mbps signal. 256 QAM is an eight-level modulation with a theoretical efficiency of 8 bit/s/Hz. Thus, the symbol rate is 43.75 Msps (megasymbols per second).

Common forms of QAM in common use are 4 QAM (QPSK), 16 QAM, 64 QAM, 128 QAM, and 256 QAM. The higher-order QAM modulation schemes are widely used in high data rate wireless systems. Such modulations are also widely used in wired applications. For example, 64 QAM and 256 QAM are often used in digital cable television transmissions and cable modems.

To demonstrate the implementation of a QAM modulator, consider the block diagram shown in Figure 3.14. The digital baseband signal is split into two independent data streams, each of which is half the bit rate of the original signal. Each half-rate data stream is then amplitude modulated, depending on the modulation level being employed. One channel (the in-phase I channel) is then multiplied by the carrier signal, while the other channel (the quadrature Q channel) is multiplied by the same signal, but with a 90° phase shift. The two signals are then added and sent to the rest of the radio circuitry for transmission.

It should be noted that both the I and Q channels are pulse-shaped via a baseband lowpass filter. This filter is necessary because the ideal square-edged digital pulses in the time domain require, in theory, infinite bandwidth to transmit in the frequency domain. Therefore, it is practical to shape the digital pulses by smoothing their edges via filtering and thus limiting the bandwidth required to transmit the pulses.

A raised-cosine filter is used for this purpose. The filter is often called a Nyquist filter, since this pulse-shaping helps minimizing intersymbol interference (ISI), as described by Nyquist theory. A key parameter of the raised-cosine filter is α, the filter roll-off factor or excess bandwidth of the filter, beyond the ideal

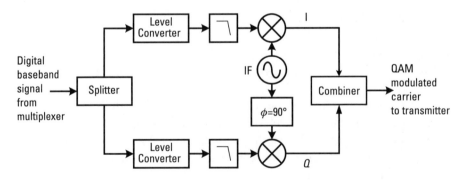

Figure 3.14 Conceptual block diagram of a QAM modulator.

bandwidth of the signal. The selection of α is a trade-off between the bandwidth required to transmit the signal and the complexity of the filtering. The bandwidth of the modulated signal BW is related to both the symbol rate F_S and α by:

$$BW = F_S \times (1 + \alpha) \tag{3.3}$$

Given a fixed modulation, it can be seen that to maximize data throughput in a fixed bandwidth, the value of α has to be minimized, creating an increasingly more complex filter. At the limit of $\alpha = 0$, an ideal "brick wall" filter is required. In practice, systems usually select a value for α of between 0.15 and 0.5 as a trade-off between throughput and implementation complexity.

It should be noted that the output signal, which now contains the initial baseband signal modulated onto a carrier, is now at a fixed frequency determined by the modulator oscillator. This is known as the intermediate frequency (IF). It is called intermediate since it is not usually the final transmission frequency, and one or more additional frequency upconversions will take place elsewhere in the radio. Some systems, however, do modulate the digital data directly onto the final transmission frequency. In such cases this is called direct modulation.

3.4.4 Demodulator

A demodulator is used to extract the desired transmitted data from a modulated signal, removing as much as possible any corruption due to analog effects of the transmission channel. Since it is closely related to the modulator, the two are often implemented in tandem, hence forming a modem (modulator – demodulator).

The simplest type of demodulator is an envelope detector. Here a diode is used to simply detect the envelope of an incoming modulated signal. This is useful for simple nonconstant-envelope modulation schemes such as ASK or OOK, where the data information is simply carried in the amplitude of the signal. For more complex forms of modulation, such as PSK and QAM, a coherent (synchronous) demodulator is required that extracts timing information from the incoming modulated signal and uses this in the demodulation process.

A block diagram of a QAM demodulator is shown in Figure 3.15. Comparison shows that it is closely related to the QAM modulator (Figure 3.14), but with two key differences. The first difference is that the demodulator includes a carrier recovery circuit, which is used to extract a stable carrier frequency from the modulated signal. This is used to develop stable I and Q channels. This carrier recovery circuit is often implemented as a Costas loop circuit, which is an extended phased locked loop configuration that enables the suppressed carrier to be extracted. The second difference is that the QAM demodulator includes a symbol clock recovery circuit, used to extract the symbol rate frequency and

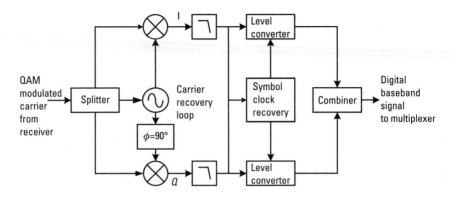

Figure 3.15 Conceptual block diagram of a QAM demodulator.

enable the level converters to sample the I and Q channels at the symbol rate and yield the final baseband signal.

One problem frequently encountered with wireless links is multipath fading, whereby echoes of the wanted signal arrive at the receiver, delayed in time and phase, and interfere with and possibly cancel out the wanted signal. Adaptive equalization is widely used in coherent systems such as PSK and QAM demodulators to compensate for this multipath distortion and, in particular, to cancel intersymbol interference (ISI) caused by signal dispersion across the channel.

Adaptive equalization is an automatically changing filter that adapts to time-varying properties of the communication channel. Often referred to as an adaptive transversal equalizer (ATE), the filter consists of a multiple-tapped delay line that can analyze the incoming signal in time and provide statistical weights for each isolated section of the signal. Since signal energy dispersion can be such that interference can arrive before or after the wanted pulse, inputs are isolated into sections both before and after the main symbol. An effective ATE therefore usually consists of both a decision feedback equalizer (DFE) and a feed forward equalizer (FFE). Generally, the more taps available in the equalizer, the higher the multipath and ISI mitigation, and the better the performance of the radio system.

Since so much of the modulator and demodulator (modem) relies on digital signal processing, the modem is usually implemented in a structured or programmable signal processing device. Since the modem functions work in serial with multiplexer, encoder, and framer functions described earlier, usually all of these features are implemented in a single ASIC or FPGA device.

3.4.5 Transmitter

The purpose of a transmitter is to upconvert the modulated IF carrier to the final transmission frequency with sufficient amplification and power to transmit the signal. A block diagram of a typical transmitter is shown in Figure 3.16.

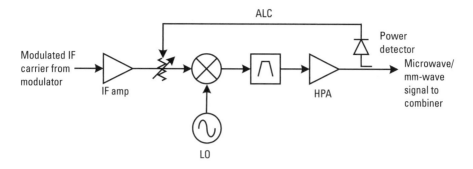

Figure 3.16 Block diagram of typical microwave/millimeter-wave transmitter.

The modulated IF signal from the modulator is first amplified and then passed to a mixer for upconversion from the IF carrier frequency to the final transmission frequency. Often, several such upconversion stages are used, depending on the application. The final transmission frequency is determined by a local oscillator (LO), which also sets the tuning range and frequency step size of the final carrier frequency.

For high-level modulation schemes, the performance of the transmitter LO is critical to the overall performance of the system. The LO needs to be synthesized, or phase locked, to a high stability source. This is usually implemented as a voltage controlled oscillator (VCO) locked to a highly stable low frequency crystal oscillator within a phase locked loop (PLL) architecture. If the LO is not synthesized, it would drift in frequency, resulting in an unstable transmitter signal. This may be acceptable for nonfrequency or phase-dependent modulations (e.g., OOK), but for frequency and phase dependent modulations, frequency stability is necessary. An important parameter of the synthesized LO is phase noise. Ideally, an oscillator would be an infinitely narrow impulse signal at a single, discrete frequency. In practice, the oscillator will be a signal distributed across a narrow spread of frequencies, around the required oscillator frequency. The distribution of the noise contained in the frequencies offset from the center frequency is known as phase noise. This has a degrading effect on higher modulation systems.

Figure 3.17 shows this degrading effect on a 256 QAM system. The addition of LO phase noise will cause variations in the phase of the modulated signal, which can be seen by the individual constellation points smearing in a radial pattern. This is particularly evident at the outer edges of the constellation, where the angular uncertainties are more noticeable. This angular noise makes it more difficult for the demodulator to correctly identify the individual data symbols. For a high-order modulation scheme like 256 QAM, the constellation points are close together and so the decision threshold for the demodulator is very small. Therefore, the smearing at the outer edges of the constellation in Figure 3.17 will cause the system to run with a noticeable bit error rate (BER). For lower order

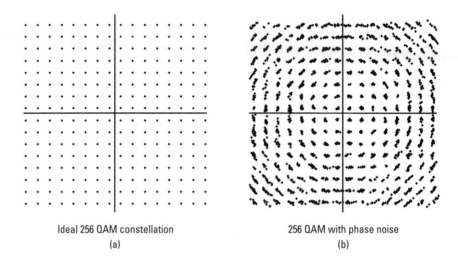

Ideal 256 QAM constellation   256 QAM with phase noise
(a)   (b)

Figure 3.17 Ideal 256 QAM (a) constellation plot and (b) practical plot with added phase noise.

modulation schemes, the spacing between constellation points is greater and so higher levels of phase noise can be tolerated.

After the mixer, a bandpass filter is used to pass the wanted upconverted microwave or millimeter-wave carrier frequency, and to reject any harmonics or mixer intermodulation products. The signal is then amplified ready for transmission. Usually two or more amplifiers are used, often referred to as the medium power amplifier (MPA) and the high power amplifier (HPA). For constant envelope modulations (e.g., FSK), the power amplifiers can be allowed to amplify right up to their maximum saturation limit. However, for other modulations, particularly higher-order QAM, the amplifiers need to be backed off from their maximum saturated levels and operate in their linear regions. This is because the higher-order QAM levels need to be able to determine between small amplitude changes, and so high levels of amplifier linearity are required. Amplifier nonlinearity will have a similar effect to that shown in Figure 3.17 for phase noise, except that amplitude variations (smearing towards and away from the original of the constellation) will be seen. It is common to see HPAs backed off up to 10 dB below their maximum power levels to provide sufficient linearity for 256 QAM transmission. Thus, QAM transmitters utilize class-A amplifiers, which are very inefficient and wasteful of power, but are necessary to preserve the fidelity required for QAM transmission. Some transmitter architectures employ predistorters to deliberately alter signals in such a way that compensates for the HPA's nonlinearity. Other systems use linearizers to sample the added HPA distortion and feed back the signals in such a way that cancellation of the distortion results. Both techniques do enable the amount of transmitter backoff to be reduced and thus the transmitter to operate more efficiency at a higher power, but often at the result of system complexity and additional tuning during the setup and commissioning of the wireless system.

The final stage in the transmitter chain is a power detector, whereby a diode is used to sample the output power and feed back an automatic level control (ALC) signal to an earlier variable attenuator or gain stage, which can enable more or less attenuation or gain, so as to keep the average output power at a required level. This is necessary for two reasons. First, as discussed above, since HPA linearity is critical for QAM modulations, a system needs to carefully control HPA operation and hence output power. Second, output power variations and levels are usually tightly specified and restricted by regulatory bodies. If output levels are exceeded, the system would violate licensing rules and risk causing interference with other nearby wireless services.

3.4.6 Receiver

The purpose of a receiver is to detect a potentially tiny wanted microwave or millimeter-wave signal from among many possible unwanted spurious signals and to downconvert this signal with a minimum of added noise so it can properly be analyzed by the demodulator. A block diagram of a typical receiver is shown in Figure 3.18.

Initially the microwave or millimeter-wave signal will be amplified by a low noise amplifier (LNA). The purpose of the LNA is to amplify the small received signal with a minimum of added noise. A key parameter is therefore the LNA's noise figure NF, which should be as small as practically possible. The amplified signal is then downconverted to a lower frequency, where it is easier to undertake more analysis of the waveform. Similar to the transmitter, the phase noise and frequency stability of the LO required to perform the downconversion are critically important to the performance of the receiver. After the mixer, the signal is bandpass-filtered to remove any spurious signal and harmonics from the mixer. Also, depending on the width of this filter, it is able to remove any unwanted signals that the LNA may have inadvertently detected and amplified. One or

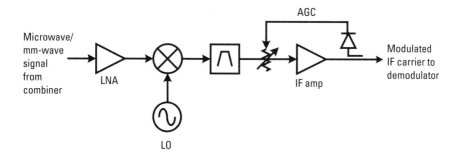

Figure 3.18 Block diagram of typical microwave/millimeter-wave receiver.

sometimes several downconversion stages are used, depending on the application. The final IF frequency is determined by the frequency of the various LOs.

After conversion to the final IF frequency, the signal is amplified before being passed onto the demodulator for processing. It is important to control the gain of this amplifier, since if the output signal is too big, it can compress the amplifier and distort the signal, or it may even overload the demodulator. Therefore, an automatic gain control (AGC) circuit, consisting of a power detector feeding back a control signal to a variable attenuator or gain device, is used to keep the IF amplifier output within the required power levels for best performance of the demodulator.

3.4.7 Combiner

A wireless system can use two antennas: one for transmitting and one for receiving, but this would be very inefficient since antennas are reciprocal devices that can both transmit and receive simultaneously. In addition, since wireless systems transmit and receive frequencies are usually very close or perhaps even at the same frequency, it makes economic and practical sense to use just one antenna. A combiner is employed to make this possible.

The purpose of the combiner is to pass the high power transmitter output signal to a single antenna for transmission, plus also to pass the usually very low level received signal gathered at the same antenna to the receiver input for signal detection. It has to do this in such a way so as to avoid the high powered transmitter signal and any intermodulation products of that signal leaking into the sensitive receiver and causing interference.

The combiner is therefore a three-port device. It can take several forms and be implemented in different ways, depending on the application. For FDD applications when both the transmitter and receiver are operational at the same time, and thus the antenna is both transmitting and receiving simultaneously, either a diplex filter or a circulator is used. For a TDD application when either the transmitter or the receiver are operational at any one time, and thus the antenna is either transmitting or receiving, but never both, a high-speed switch is employed. Key performance attributes are the transmitter to antenna insertion loss, which should be as small as possible to minimize the loss of the high power transmit signal, and the transmitter to receiver isolation, which should be as large as possible to avoid leakage of the higher power transmitter signal into the very sensitive receiver chain. For a TDD switch in a high data rate radio system, switching speed is also of critical importance. These three combiner configurations are shown in Figure 3.19.

Higher-level combining techniques that allow the outputs of several radios to be aggregated together, such as branching networking for multiple chan-

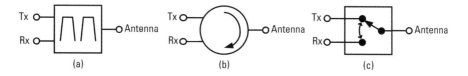

Figure 3.19 Three common forms of combiners: (a) diplex filter, (b) circulator, and (c) switch.

nel systems and combiners for cross-polarization applications, are considered in Sections 4.4.2–4.6.

3.4.8 Antenna

An antenna is a critical part for any wireless system. On transmit, an antenna converts the high powered electrical signal from the transmitter into an electro-magnetic wave and concentrates the energy into a focused beam that propagates across the wireless channel. On receive, the same antenna collects energy incident on its surface area and converts this back to an electrical signal for detection by the receiver.

There are many types of antenna in common use for different wireless applications. For long-distance communications, highly directional reflector or aperture antennas are used. For high coverage applications such as cell sites, wide beamwidths are achieved by using very different broad-beam antennas. Antennas can be large metal dishes or can be tiny planar circuit elements printed on a circuit board. Antennas can be fixed or steerable. There are two types of antenna in common use for high capacity systems. There are the parabolic antenna for high-performance, longer-distance use and the flat panel antenna, either discrete or planar, for low cost, shorter distance applications. A parabolic antenna operates by having the feed horn placed at the focal point of the parabolic curved dish, such that all reflections form a plane wavefront with all the individual rays parallel and in phase. A cross-section of such a parabolic antenna is depicted in Figure 3.20. A flat panel antenna produces a similar planar wavefront, but it usually generated by one or more smaller antennas, operated in a phased array to shape the beam.

The power radiated from or received by an antenna is expressed as the gain of the antenna G. This is quoted as gain relative to an isotropic radiator (an ideal antenna that radiates energy in all directions, that is, outwards in a sphere), and is given by:

$$G = \frac{4\pi\eta A}{\lambda^2} \tag{3.4}$$

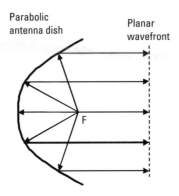

Figure 3.20 Cross-section of a parabolic antenna, showing position of focal point F.

where η is the efficiency of the antenna, λ is the wavelength of the antenna's operating frequency, and A is the area of the antenna's aperture.

For a high frequency antenna, efficiencies are typically between 50% and 60%. For a circular antenna with 55% efficiency, such as a parabolic antenna widely used at microwave and millimeter-wave frequencies, (3.4) is often simplified to give:

$$G = 17.8 + 20\log_{10}\left(f \cdot d\right) \qquad (3.5)$$

where gain G is measured in dBi, frequency is measured in gigahertz, and d, the diameter of the antenna dish, is measured in meters. It is worth noting that the gain of the antenna is directly proportional to the square of the diameter of the dish and the square of the frequency. Thus, moving up in frequency allows significant reduction in the antenna size to maintain the same gain. Also, doubling the linear dimensions of any antenna will add 6 dB of antenna gain or 12 dB of system gain improvement if both antennas at either end of the link are doubled in size.

Directivity is another parameter commonly used to describe antenna performance. Since an antenna is a passive device, it cannot amplify the signal. However, it can shape the signal to focus energy, allowing the signal to be stronger in one direction than in others. In practice, it is impossible to focus all the energy into one direction, and an antenna always has sidelobes, which emit smaller amounts of electromagnetic radiation out of the sides and back of the antenna. A directional antenna aims to maximize the energy in the main transmission lobe by minimizing the energy in these sidelobes and back lobes. Directivity is the antenna's focusing ability to maximize the main front lobe that determines the antenna's gain.

Gain and directivity are commonly specified via an antenna radiation pattern envelope (RPE). This can be plotted on either polar or rectangular coordinates, as shown in Figure 3.21. The important concept of bandwidth is illustrated. Commonly known as the half-power beamwidth (HPBW), this is the total angle of the main antenna lobe when measured at half the power of the main lobe. Bandwidth is most easily read off the rectangular coordinates simply as where the RPE is 3 dB lower than the antenna's gain. It should be noted that it includes both the positive and negative angles combined. It should also be noted that an antenna has a three-dimensional RPE and hence the two-dimensional RPE plots and beamwidth measurements may be different in the horizontal and vertical transmission planes.

For a parabolic antenna, the half-power beamwidth (measured in degrees) can be estimated from the well-known rule of thumb that says:

$$HPBW = \frac{70\lambda}{d} \tag{3.6}$$

Assuming a circular parabolic with efficiency of 55%, substituting (3.4) into (3.6) and simplifying, (3.6) simplifies to:

$$HPBW = \frac{163}{\sqrt{G}} \tag{3.7}$$

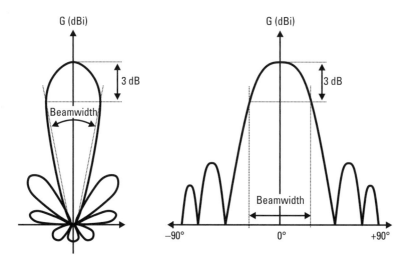

Figure 3.21 Polar and rectangular coordinate representations of the same antenna radiation pattern, showing the measurement of half-power beamwidth.

where G is the absolute gain, which can be determined from the gain in dBi (G_{dBi}) by

$$G = 10^{\frac{G_{dBi}}{10}} \tag{3.8}$$

Another important antenna characteristic is front-to-back ratio. This is the ratio of the power radiated from the main front lobe to that radiated from the back lobe of the antenna. This can be seen in Figure 3.20 where there is a small lobe extending from the back of the antenna. All antennas radiate some energy backwards out of the antenna. For high directivity applications, the front-to-back ratio needs to be maximized.

3.4.8.1 Parabolic Antennas

In practice, microwave and millimeter-wave parabolic antennas are usually constructed from spun aluminum or metalized plastic and often built with an added shroud around the edge to reduce radiation from the sides and back of the antenna. This is required to maximize the antenna gain, minimize sidelobes, and maximize the front-to-back ratio. As such, parabolic antennas are often used when system performance is critical.

3.4.8.2 Flat Panel Antennas

Flat panel antennas are often the choice for lower cost systems and consumer applications. Here the requirements are not necessarily highest performance, but are often low cost, small size, and lightweight. All of these can be readily achieved with flat panel designs. Despite these advantages, flat panel antennas have limited frequency responses, making it difficult to achieve wideband operation. For example, it is difficult to achieve constant gain and high efficiency over a broad frequency range such as the 57–66-GHz bandwidth available in the 60-GHz band.

A flat panel antenna is usually formed by multiple smaller antennas integrated together to form antenna arrays for both fixed and portable applications. By using multiple antennas, it is possible to increase the antenna gain beyond that of a single antenna. In addition, beamforming can be implemented by either switched beam arrays or phased arrays. This can be very beneficial in residential applications, where human movement can easy block or attenuate a wireless signal. A switched beam antenna array or an adaptive antenna array can be implemented to steer a signal and capture the wanted signal, making use of strong reflected signals when the main signal is attenuated. The array is required to track the signal path either continuously or periodically, depending on the stability of the link. Switched beam arrays have multiple fixed beams that can be selected to cover a given service area. They can be implemented more easily than phased arrays, which require the capability of continuously varying a progressive phase shift between the elements.

One major parameter for the performance of a flat panel array antenna is how many antenna elements are required to achieve the intended antenna gain. Also of interest is the angular resolution or beamwidth of such antennas, since this defines the number of multipath signals that the antenna sees in a scattering environment. The directivity of a linear array of antennas is given by:

$$D = \frac{4\pi}{\iint |F_n(\phi,\theta)|^2 \sin\theta \, d\theta \, d\phi} \qquad (3.9)$$

where $F_n(\phi, \theta)$ is a normalized field pattern for each element in the antenna, and ϕ and θ represent the azimuth and elevation angle, respectively. More detailed discussions on flat panel antennas can be found in [1].

3.4.8.3 Other Antenna Elements

Sometimes antennas are integrated in the wireless system and form an integral part of the RFU. On other systems, the antennas are a physically separate item that can be swapped for different sizes and types to suit the application.

For external antennas, connection between the RFU and the antenna is required. A coaxial cable feed is common at lower frequencies. Coaxial cable is a relatively low loss transmission medium, whereby the signal is propagated on a central conducting core. The signal is contained within a surrounded dielectric medium by an outer metallic conductor sheath that is usually grounded. Because such cables (and particularly their connectors) have losses of typically several decibels or so at the higher microwave and millimeter-wave frequencies, waveguide is often the preferred interconnect to antennas. Waveguide consists of a metallic tube, either circular or rectangular, that guides an electromagnetic wave with very little loss.

Finally, a radome is often used to cover an antenna. This is usually built of fabric or fiberglass, both of which have low dielectric losses. Solid radomes are usually chemically coated to help avoid water, ice, and snow buildup during harsh weather. The radome protects the antenna and, in particular, its delicate feed system from the elements. A solid radome also reduces the wind loading on an antenna considerably.

3.5 Summary

High data rate wireless systems have a variety of different architectures and physical structures to match the application in which the wireless system is being used.

Some units are compact, consumer devices for in-home residential applications. Others are physically large units mounted either indoors, outdoors, or often split between the two for high-performance industrial applications.

High data rate systems operate with a variety of over-the-air wireless protocols, some of which are proprietary and some of which are standards based. High data rate systems also operate with a variety of high data rate line protocols. These are always standards-based, conforming to a recognized format so that equipment can be interconnected with other devices in a network. Typical examples of such high data rate line protocols are SDH/SONET and gigabit Ethernet.

A high data rate wireless system can be broken down into a number of generic subassemblies, each of which contributes necessary functionality to the system. These include the power supply, data interface and multiplexer, modulator, demodulator, transmitter, receiver, combiner, and antenna. Each has key characteristics and performance attributes that must be correctly designed and implemented for optimal operation of the overall wireless system.

Reference

[1]   Huang, K-. C., and D. J. Edwards, *Millimeter Wave Antennas for Gigabit Wireless Communications*, New York: John Wiley & Sons, 2008.

4

Multi-Gigabit Microwave Radios

4.1 Introduction

Wireless systems are widely used to distribute voice and data, supporting services such as broadcast radio, GPS, cellular, WiFi, WiMAX, high-definition audiovisual, and a host of other applications. The voice and/or data carrying capacity of each technology depends primarily on the amount of radio spectrum in the frequency band in which the service operates and, in particular, the bandwidths allocated by regulators for that service. Generally, the higher the frequency of the service, the more spectrum and bandwidth are available to support that service, and so the greater the data rate that the technology can support.

The microwave bands of 6 to 40 GHz can support full-duplex data rates of many hundreds of megabits per second. There are multiple systems and applications supported in these frequency bands. Many are low data rate, such as radars and speed detectors. Others are higher data rate, such as satellite services and TV distribution. The highest commercial data rates achieved in the microwave bands are for terrestrial fixed wireless radio systems where data rates in excess of 1 Gbps can be achieved. These are widely used for providing high-speed PTP connectivity between cell sites (cellular backhaul), and in private networks connecting together offices, campuses, and businesses (enterprise connectivity). In 2009, over 1 million such PTP microwave radios were sold worldwide.

Practically realizable data rates for 6–40-GHz microwave radios are limited by the bandwidths allocated to such services. In general, these are a maximum of 50 MHz in the United States and 56 MHz in the rest of the world. Up until a few years ago, the maximum throughput supported by commercial microwave equipment was 311 Mbps, achieved by systems employing high-order 128 QAM modulation to squeeze two 155-Mbps OC3/STM1 data streams into these maximum channel sizes. With the recent push towards the more flexible Ethernet data standards, commercial systems are widely available that use 256 QAM to achieve data rates of around 350 Mbps and higher. To achieve higher data rates, systems are being architected using dual configurations and parallel transmissions to achieve double the throughput. Such systems can use adjacent channels, or can transmit within the same channel on different polarizations, or a combination of both to reuse frequencies for dual and quad data streams, doubling and even quadrupling the effective data throughput. Such configurations can yield full-duplex data rates in excess of 1 Gbps.

Since microwave radio systems have been commercial available for over 30 years, many texts have been written covering these devices, this technology, and how it is used (see, for example, [1]). This chapter focuses on how such microwave radio systems can be used to deliver high-speed wireless capacity at data rates of 1 Gbps and beyond.

It is worth repeating the earlier note that strictly speaking, the term microwave refers to the SHF frequencies of between 3 and 30 GHz (see Chapter 1). However, in this chapter, the bands 6 GHz to 40 GHz are referred to as microwave. This is because these bands conveniently have similar characteristics and physical properties, are made available and are managed in a similar way by global regulators and licensers, and are used for similar applications. Therefore, it is convenient to groups these services together. Conventionally, such technologies are referred to as the 6–40-GHz microwave bands.

4.2   Characteristics of the Microwave Bands

4.2.1   Frequency Bands and Channel Sizes

The 6–40-GHz microwave bands are widely available around the world for high data rate wireless communications. This 6–40-GHz spectrum is divided into specific bands that are made available by national regulators in individual countries. There are often small regional changes in band plans, but most countries follow the allocations set by the United States or Europe. In the United States, the frequency bands are set up, administered, and managed by the FCC. In Europe, the frequency bands are set up by CEPT, technical rules for their use are defined by ETSI, and the bands are administered and managed by individual countries national regulators.

Table 4.1 shows the major microwave frequency bands available in the United States and Europe for terrestrial fixed link applications. Also shown are the channel sizes that regulators have allowed in each of these bands.

In the United States, the 6-GHz and 11-GHz bands are widely used for terrestrial wireless transmissions. The two bands have very strict technical restrictions on their use. For example, in the 6-GHz band, a minimum antenna size of 6 feet (2m) is required, making it very useful for long-distance transmissions. The 18- and 23-GHz bands have much less severe restrictions on their use. These are widely used for general purpose communications, especially for higher data rate applications because of the larger channel sizes permitted. In the 18-GHz band, for example, channel sizes of 40 MHz, 50 MHz, and 80 MHz are permitted. In the 23-GHz band, larger channel sizes of 40 MHz and 50 MHz are permitted. The 28-GHz and 38-GHz bands are managed very differently to the lower bands. The frequencies have been auctioned and raised significant money for the FCC. After many mergers, acquisitions, and legal battles, the frequencies are now held by just a few companies. These owners do not permit general usage.

Table 4.1
Major Microwave Frequency Bands Available for Terrestrial Wireless Transmissions in the United States and Europe

Frequency Band	United States		Europe	
	Frequency Limits	Bandwidths	Frequency Limits	Bandwidths
Lower 6 GHz	5.925–6.425 GHz	400 kHz–30 MHz	5.925–6.425 GHz	29.65 and 59.3 MHz
Upper 6 GHz	6.525–6.875 GHz	400 kHz–10 MHz	6.525–6.875 GHz	3.5–60 MHz
7 GHz			7.125–7.725 GHz	1.75–56 MHz
8 GHz			7.725–8.5 GHz	7–60 MHz
10 GHz	10.55–10.68 GHz	400 kHz–5 MHz	10.0–10.68 GHz	3.5–56 MHz
11 GHz	10.7–11.7 GHz	1.25–40 MHz	10.7–11.7 GHz	40 MHz
13 GHz			12.75–13.25 GHz	1.75–56 MHz
15 GHz			14.5–15.35 GHz	1.75–56 MHz
18 GHz	17.1–19.7 GHz	1.25–80 MHz	17.7–19.7 GHz	1.75–55 MHz
23 GHz	21.2–23.6 GHz	2.5–50 MHz	22.0–23.6 GHz	3.5–56 MHz
26 GHz			24.5–26.5 GHz	3.5–56 MHz
28 GHz			27.5–29.5 GHz	3.5–56 MHz
31 GHz	27.5–31.3 GHz	Block allocation	31.0–31.3 GHz	3.5–56 MHz
32 GHz			31.8–33.4 GHz	3.5–56 MHz
38 GHz	38.6–40.0 GHz	Block allocation	37.0–39.5 GHz	3.5–56 MHz

However, they are able to use the bands with minimal restrictions, and so can use high channel sizes and thus support very high data rate transmissions.

Therefore, within the United States, the following frequency bands and channels are favorable for high capacity wireless communications:

- 80-MHz channel at 18 GHz;
- 50-MHz channel at 18 and 23 GHz;
- 40-MHz channel at 11, 18, and 23 GHz.

Although there is this wide 80-MHz channel, its availability is limited and channel allocations are quickly used up. Given the single occurrence, there is not enough channel slots to support a market for 80-MHz equipment. For this reason, operators usually use European 56-MHz equipment when they have access to this particular channel.

In Europe, the situation is very different. There are a much larger number of available frequency bands, all of which are generally available for anyone wishing to purchase a license. Rules governing each band are generally consistent, making the situation much simpler than in the United States. In the majority of frequency bands, 56-MHz channels are available. A few of the lower frequency bands can also support the slightly higher 60-MHz channels, but like the sole 80-MHz channel in the United States, access to this 60-MHz channel is limited.

For these reasons, the widest generally available channel sizes for high data rate microwave communications is 50 MHz in the United States and 56 MHz in Europe and most of the rest of the world. Throughout the remainder of this book, these are considered the widest microwave channels available.

4.2.2   Rules and Regulations

Rules and regulations for approving, operating, and managing microwave wireless equipment are very different in the United States and in Europe.

4.2.2.1   United States

In the United States, all telecommunication devices under the jurisdiction of the FCC are managed under "Title 47 of the Code of Federal Regulations" (47 CFR). These cover a wide range of rules and regulations from marketing and certifying of devices up to technical performance requirements and permitted applications. Of relevance to high data rate microwave communication systems are Part 15 (47 CFR §15, [2]) and Part 101 (47 CFR §101, [3]). Part 15 is often misunderstood to apply to just unlicensed devices, but it does in fact apply to all intentional and unintentional radiators, and thus covers all radio frequency devices, licensed and unlicensed. Part 101 is applicable to fixed wireless devices.

There are many other technical rules and regulations that are applicable to other services (commercial radio, amateur radio, satellite communications, aviation services). Since these services are not able to support gigabit per second communications, they are not considered further here.

For high capacity wireless devices, Part 101 covers just a few critical technical parameters such as frequency limits, allowed modulation schemes, radiated power, spurious emissions, bandwidths, and frequency tolerance. Part 15 covers the allowances for unwanted radio emissions. Any fixed wireless equipment offered for sale in the United States needs to be certified to the relevant technical requirements in these two parts. Depending on the equipment, this can be through verification (self-certification) or by formal certification by the FCC or an FCC-approved Telecommunications Certification Body (TCB).

Although not mandated by the FCC, microwave wireless equipment can also be certified to various other standards, depending on the application and operational environment. For example, to operate in telecommunications environments, Network Equipment Building System (NEBS) safety certification is often required. There are three levels of increasing protection, from basic protection from equipment hazards, right up to fire suppression and vibration (earthquake) resistance. Another safety specification often met by wireless equipment is UL 60950.

4.2.2.2 Europe

In Europe, a completely different set of rules and regulations apply. These are completely unrelated to those in place in the United States and generally far more detailed and thorough. The technical rules applied to microwave fixed links are defined by ETSI specification EN 302 217 [4]. This is actually a set of multipart specifications that each deal with a particular aspect of fixed wireless equipment and characteristics. There are six subparts, of which the titles relevant to high data rate microwave radios are:

- *Part 1:* Overview and system-independent common characteristics;

- *Part 2-2:* Harmonized EN covering essential requirements of Article 3.2 of R&TTE Directive for digital systems operating in frequency bands where frequency coordination is applied;

- *Part 4-2:* Harmonized EN covering essential requirements of Article 3.2 of R&TTE Directive for antennas.

Part 2-2 contains the technical specifications for microwave wireless equipment. This specification, often referred to as RF type approval, details a large range of technical requirements and test methodologies for RF and microwave equipment up to 55 GHz, including:

- Transmitter requirements:
  - Output power;
  - Output power tolerance;
  - Transmitter spectrum mask;
  - Transmitter control [e.g., Automatic Transmit Power Control (ATPC) and Remote Transmit Power Control (RTPC)];
  - Spurious emissions;
  - Frequency tolerance.
- Receiver requirements:
  - Spurious emissions;
  - Bit error rate (BER) as a function of receiver input signal level (RSL);
  - Cochannel interference sensitivity;
  - Adjacent channel interference sensitivity;
  - CW spurious interference.

Also, references to other significant specifications, such as environmental conditions and profiles, are detailed.

Part 4-2 is a similar technical specification for antenna characteristics, which provides requirements for the radiation pattern envelope (RPE) and cross-polar discrimination (XPD) for antennas in the microwave bands considered here and also the millimeter-wave bands considered in later chapters.

In Europe, wireless systems are also required to be certified to a number of other specifications. For example, electromagnetic compatibility (EMC) conformance is mandatory. EMC is governed by ETSI specification EN 301 489 [5], which contains 32 parts covering different categories of wireless equipment. The part relevant for high data rate microwave wireless equipment is titled: Part 4: Specific conditions for fixed radio links and ancillary equipment and services.

This specification defines the requirements for both radiated and conducted emissions and immunity. (Note that in the United States, only compliance to radiated emissions is required.) EN 301 489-4 covers the following:

- Radiated emissions from the enclosure;
- Conducted emissions:
  - From the AC and/or DC port and signal ports;
  - Harmonic current emissions from the AC port (if applicable);
  - Voltage fluctuations and flicker from the AC port (if applicable).
- Radiated immunity:
  - RF field onto enclosure;

- Electrostatic discharge (ESD) onto enclosure.
- Conducted immunity:
  - RF interference onto AC and/or DC port and all signal and control ports;
  - Fast transients onto AC and/or DC port and all signal and control ports;
  - Transients and surges onto DC ports (if equipment used in mobile applications);
  - Voltage dips and interruptions into the AC port (if applicable);
  - Surges into the AC port (if applicable).

The final area where compliance is required is safety. The relevant specification for microwave radio equipment is EN 60950-1 [6], which covers general information technology equipment. This is a very wide ranging specification with numerous requirements based on the original International Electrotechnical Commission (IEC) spec IEC 60950. (Outside Europe, many countries have their own version of this specification, for example, UL 60950 in the United States, CSA 60950 in Canada and AS/NZS 60950 for Australia and New Zealand, all of which are based on the same IEC 60950 base specification.)

Before equipment can be offered for sale in Europe, it needs to meet the essential requirements of the European Commission's Radio and Telecommunications Terminal Equipment (R&TTE) Directive. These essential requirements cover the health and safety of the user and others, electromagnetic compatibility, and effective use of the radio spectrum. All approved wireless devices need to carry a CE mark before they can be offered for sale in any European country. A CE mark is obtained by submitting a Technical Construction File (TCF) to a certified Notified Body, and receiving a Declaration of Conformity (DoC) from that Notified Body. The TCF requires compliance to the three necessary specifications detailed above (RF type approval, EMC and safety). Once approval is obtained, a CE mark can be affixed to the device, and it can be freely marketed and sold across Europe.

4.2.3 Licensing

As discussed in Section 1.2, the radio spectrum is a finite resource. Demand greatly exceeds supply in many frequency bands and in many well populated areas. This means that the use of frequencies needs careful planning in order to make the best use of the available spectrum, as well as ensuring that minimum interference is caused to authorized radio users. The basis for this planning is through the issue of licenses by individual country's national frequency managers. Installation or use of a radio service without a valid license is usually an offense under the county's wireless communication laws, the exceptions generally being equipment that is categorized as "license-exempt" or equipment used by the national government.

Just about all high capacity microwave links require licensing. A license is a legal authority allowing the licensee to install and operate wireless equipment in a defined way within a defined geography. Once a license is issued for a wireless link, no other user or service can operate at the same frequencies within the same geographic area, thus ensuring that the risk of interference to that link is minimized. This is especially important for high data rate links since the link owner and/or operator is likely to be generating significant revenue from the services enabled by the high capacity link and does not want to risk downtime and the resulting loss of revenue that may result.

Licensing processes vary country by country and application by application. In the United States, licensing for commercial wireless equipment is managed by the FCC. The process is well defined and documented. For the licensing of fixed link microwave systems, for example, the licensee first has to register with the FCC through their Commission Registration System (CORES) and receive an FCC Registration Number (FRN). This number is uniquely used to identify the licensee during all transactions with the FCC. Once registered, the licensee has to provide all information about the proposed link using Form 601: Application for Wireless Telecommunications Bureau Radio Service Allocation. This can be done online via the FCC's Universal Licensing System (ULS). Information about the positioning of each end of the link (longitude, latitude, antenna height, link distance) and information about the wireless equipment (output power, antenna gain, EIRP, emissions designators) are required. A fee is assessed for each call sign (location), which is currently set at $560. A separate frequency coordinator is also often used to undertake interference calculations and frequency assignments to determine actual frequencies within the wider frequency bands. Typically, this may cost $2,000 per license, plus adding several weeks to the overall application time.

In the United Kingdom, the process for licensing microwave equipment is similar. Ofcom, the U.K. national regulator, manages fixed link licensing via form OfW85, which issues licenses under the terms of the United Kingdom's Wireless Telegraphy Act of 2006. Many other national regulators manage the licensing process in similar ways.

Where countries generally differ is in the pricing for the license fee. In the United States, the FCC-established license fee for fixed link microwave radios is a set amount. It is not dependent on the frequency or bandwidth of the wireless transmission or the data throughput of the link. However, most other countries have pricing schedules that vary with frequency, occupied bandwidth, or data throughput. Such varying schedules can be in place for a variety of technical or political reasons. Usually, varying pricing schedules are implemented to promote more efficient use of the spectrum and to discourage use of unfavored frequency bands. In the United Kingdom, for example, Ofcom has initiated a license pricing structure that favors higher frequencies and smaller channel sizes to discourage inefficient use of the more congested, lower frequency bands and promote more spectrally efficient, higher modulation wireless products.

4.2.4 System Characteristics

In addition to there being numerous available frequency bands, with wide channel bandwidths available for high data rate license-protected operation at reasonable costs, microwave radios provide a reliable, frequency-friendly technology that permits high data rates over long transmission distances.

As shown in Chapter 7, microwave radios are regularly installed with 99.999% system availability (i.e., the radio will operate for 99.999% of the time, or only be down for 0.001% of the time or approximately 5 minutes per year). Microwave radio equipment accomplishes this by using robust equipment architectures, forward error correction, and other bit error mitigation schemes. Equipment can be provisioned in protected and path diversity configurations to maximize uptime and reception quality. These proven features and techniques are widely employed where reliability is absolutely critical to delivering uninterrupted service.

In addition, microwave radio links can provide long-distance transmissions, covering distances of over 50 miles (80 km) for some of the lower frequency bands. The lower frequency radios typically have higher transmitter output power and are usually operated with large antennas, aiding their long-distance capabilities.

Finally, significant thought has been placed in allocating and channelizing frequency bands to promote spectrum efficiency and frequency reuse. Since the 6–40-GHz frequencies used for fixed links microwave radios are separate and noninterfering with most other wireless services (particularly the 3G and 4G cellular and WiMAX bands), high capacity microwave radios can be used to complement these services. Network planners are able to preserve RF channels for base station operation and eliminate any possibility of interference with the lower frequency service's base station or any other operator delivering services in that area.

4.3 Market Applications

The biggest application for high-speed microwave radios is PTP backhaul, which is mainly used in two applications: mobile or cellular backhaul and private network enterprise connections.

4.3.1   Mobile Backhaul

Mobile backhaul is for connecting together cell sites and cellular base stations. PTP wireless has been employed in mobile network infrastructure applications for connecting together cell sites for more than 30 years. At the end of 2009, estimates show there were approximately 4 million cell sites in the world, and that approximately 50% of these are connected by PTP wireless. Use is particularly strong in Europe where over 80% of all cell sites have wireless backhaul. With the rapid progression of wireless standards, requiring more backhaul capacity and improvements in transport protocols, cell sites often support several links satisfying legacy, current, and sometimes future services. With the growth of WiMAX networks, PTP wireless is proving to be a suitable backhaul for these cellular-like networks.

High capacity microwave is especially advantageous for supporting WiMAX, LTE, and future 4G mobile backhaul networks. A typical cellular network utilizing high data capacity microwave radios for mobile backhaul is shown in Figure 4.1. Current implementations of both LTE and WiMAX have peak theoretical data rates approaching 100 Mbps, and future 4G next generation networks can be as high as 1 Gbps. Verizon Wireless is planning to commission its LTE network in the United States in 2012. Verizon Wireless has indicated that by 2015 its average U.S. cell site will require 200 Mbps of backhaul, core cell sites will require 400 Mbps of backhaul, and wireless hub sites will require a full 1 Gbps of backhaul. Further back in the network, fiber connections to the mobile switching center (MSC) will be 10 Gbps for the smaller transport providers and 100 Gbps for the core routers [7]. Technical analysts predict that data

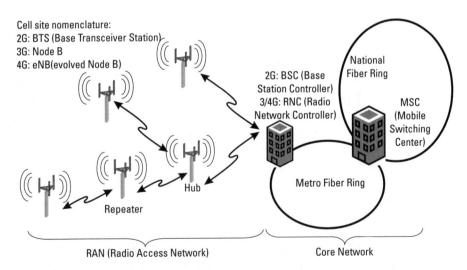

Figure 4.1 Typical cellular network architecture, showing system component nomenclature for 2G, 3G, and 4G technologies.

rates will be higher, with average backhaul capacity in markets with high data traffic expected to reach 300 Mbps by 2012 [8].

In addition to the need for gigabit per second wireless connectivity, the 4G platform is built around Internet Protocol (IP), a data-oriented transmission protocol. This is very different to the TDM voice or circuit-switched data systems of existing and older standards. IP traffic is very dynamic in terms of capacity and bandwidth, characteristics that are difficult to support in a TDM-only environment. As such, new microwave radio installs will be required to support 4G services, overlaying older networks.

4.3.2 Enterprise Connectivity

Enterprise connectivity refers to the connecting together of businesses, schools, hospitals, campus buildings, and any other enterprise. Often it is simply too prohibitive to commission fiber in commercial environments, so wireless forms a convenient method to connect users. Since enterprise connections are usually extending businesses local area networks (LANs), this application is often called LAN Extension.

High-speed microwave radios are particularly applicable for enterprise connections because LANs usually operate at 100-Mbps Fast Ethernet or 1-Gbps GbE speeds. Therefore, having a wireless connection operate at these speeds avoids having choke points in the network.

A typical high-speed enterprise connection is shown in Figure 4.2.

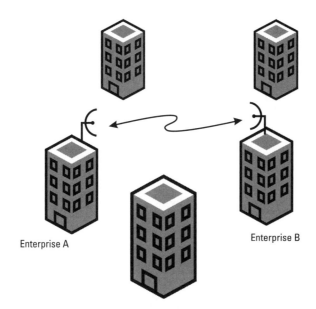

Figure 4.2 Typical enterprise architecture.

4.4   Methods for Achieving Gigabit Per Second Data Rates

4.4.1   Single Channel Transmission

To support high data rates in the 6–40-GHz microwave bands, wireless devices need to squeeze the required user data into the available regulatory channels for the required transmission frequency band. Up until about 2005, the highest data rates supported by microwave radios was 2 x STM1/OC-3 or 311 Mbps. Such systems used 128 QAM modulation to compress the data into the widest 50- or 56-MHz channel sizes available. With the more recent migration towards Ethernet transmission, where data rates can be dynamic and no longer restricted to the granular levels of PDH, SDH, and other line protocols, higher data rates and modulation schemes are being routinely deployed. Newer microwave radios can now support FDD data rates of approximately 350 Mbps and sometimes higher in a single microwave channel.

4.4.1.1   Typical System

Consider a typical microwave radio operating with 256 QAM modulation in a 56-MHz channel. A key regulatory parameter is the transmission mask, which requires that the occupied bandwidth of the channel be contained within the 56-MHz space to avoid any leakage into adjacent channels that would cause interference. To meet this mask under all practical operating conditions (for example, fluctuations in temperature, variations in synthesizer stability, manufacturing, and component variations), some margin to this mask must be allowed. Allowing a 5% design margin, a high data rate system might be designed with an actual 53-MHz occupied bandwidth to ensure compliance with the 56-MHz mask. As shown in Chapter 3, the symbol rate of a wireless device is related to the occupied bandwidth and the roll-off factor of the modem filter α. Given a typical α of 15%, (3.3) shows this radio will have a symbol rate of 46 Msps. Using (3.2) to convert this to bit rate, using a 256 QAM modulation ($n = 8$ b/s/Hz), the link would be expected to have an approximate capacity of 370 Mbps. This throughput represents the over-the-air data rate (sometimes called gross data rate) between the two wireless devices. It includes not only the customer data but also overhead data such as FEC, network management, and any auxiliary traffic. Typically such traffic adds 5% to 10% overhead, depending on FEC strength and other implementation choices, meaning that about 350 Mbps of customer available throughput can be achieved. Depending on design margins, modem parameters, and implementation efficiencies, this number will vary between equipment manufacturers. More aggressive implementations are able to achieve close to 400 Mbps of customer available data throughput under the operating conditions outlined above.

A representation of a split mount microwave radio, operating in a single 56-MHz channel and achieving 350-Mbps throughput is shown in Figure 4.3.

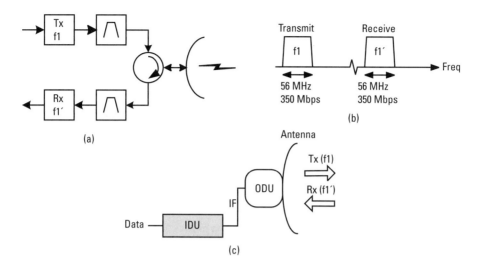

Figure 4.3 Representation of a 56-MHz channel, 350-Mbps data rate device. (a) Block diagram, (b) frequency plan, and (c) typical physical representation.

A photograph of a commercially available microwave radio capable of exceeding this performance is shown in Figure 4.4. This radio is available in the full 6–40-GHz microwave range and is able to provide 360 Mbps of robust customer available data capacity in a 56-MHz channel. Using lighter FEC coding, this data rate can be extended beyond 380 Mbps. By employing a number of the Ethernet compression techniques discussed later, rates in excess of 450 Mbps can be achieved.

4.4.1.2 Interpretation of Data Rates

It is useful to comment at this stage on quoted data rates of microwave wireless systems. There are various ways to characterize data capacity, and given the competition in the marketplace, there is temptation to quote throughputs as high as possible to create the impression that one wireless system is faster than another. This is often not the case, as data throughput is primarily a function of channel bandwidth, the system architecture used to utilize that channel, and the operating parameters of the radio. Since channel sizes are fixed, the only options available to manufacturers to differentiate their data throughputs are dependent on the choice of only a few radio design parameters. These are generally within narrow ranges, depending on how aggressively or conservatively vendors approach their system design. As such, all practical wireless systems, given the same operating parameters (modulation and channel size) offer data rates that are relatively close to one another. Large differences in quoted data rates are therefore usually an indication of special operational conditions or measurement circumstances, rather than better or worse performing equipment.

Figure 4.4 Photograph of a microwave radio able to provide 360 Mbps of full customer available throughput in a 56-MHz channel. (*Source:* Aviat Networks, 2010. Reprinted with permission.)

Up until several years ago, when all high capacity wireless systems were TDM based, there was no debate about throughputs. TDM data rates are very granular, and radios were either able to transmit one or two OC-3/STM1 data streams, or not. With the move to the more flexible and dynamic Ethernet protocol, things are now different. Consider the above example of a 256 QAM radio operating in a 56-MHz channel and producing 370 Mbps of over-the-air data and 350 Mbps of usable customer data. Vendors have been seen marketing such products as Gigabit Ethernet radios (referring to the device's 1,000baseT Gigabit Ethernet interface), 740-Mbps radios (referring to the link's total aggregate gross data rate), or 370-Mbps radios (referring to the link's over-the-air FDD gross data rate). None of these is strictly incorrect, but each is misleading if the frame of reference is not specified. More correctly, the link should be referred to as having 350-Mbps throughput, since this is the customer available data rate. However, even this can be misleading, as the amount of overhead (FEC, network management, and auxiliary traffic) can be varied to trade-off data capacity throughput with link performance. It can be seen that there are numerous ways to quote the performance of a single wireless configuration, depending on the intended end customer and how aggressive the manufacturer wants to be with the marketing of the product.

4.4.1.3   Interpretation of Ethernet Data Rates

To make this quantifying of usable data throughput even more difficult, Ethernet data throughputs can be quoted different ways, depending on whether they are Layer 1 or Layer 2 in the Open System Interconnection (OSI) model. The OSI model is a layered framework describing a communications network. Wireless terminals operate at Layer 1 over-the-air, which is the physical (PHY) layer used to describe the bit-level operation of a device. However, devices that incorporate a framing device and/or an addressing structure such as a switch operate at Layer 2, known as the data link layer. Most high capacity Ethernet radios contain an integrated switch (see Section 3.4.2) and can therefore be viewed as both Layer 1 and Layer 2 devices. The data throughput parameters specified can vary significantly depending upon which Layer is being considered.

Consider a full Ethernet frame, as defined by IEEE 802.3 [9]. A simplified version of this is shown in Figure 4.5. Each frame carries various pieces of information in addition to the actual data payload. Three sections of the Ethernet frame are of interest here: the interframe gap (the spacing between consecutive frames), the preamble (an introduction to the Ethernet frame), and the start of frame delimiter (a single byte that clarifies when the frame starts). Together these functions are allocated 20 bytes at the start of each Ethernet frame. Following this is what is usually referred to as the Ethernet frame (or more accurately, the Ethernet Type II Frame), which is typically 64 to 1,518 bytes long. This contains the actual transmitted payload, plus information about the source and destination address, and a small error correction block.

In a radio system that incorporates an integrated Layer 2 switch, only the 64 to 1,518 byte Ethernet frame is transmitted over the link. The 20 bytes of introductory information (interframe gap, preamble, and frame delimiter) are stripped from the incoming user data stream and then reinserted back into the data stream at the far end network interface by the radio equipment. Thus, the user available Layer 2 data that is transmitted over the link is always less than the Layer 1 line rate, by an amount described by:

$$\text{Layer 1 data rate} = \text{Layer 2 data rate} \times \left(\frac{\text{Full frame size}}{\text{Ethernet frame size}} \right) \qquad (4.1)$$

Now consider again the earlier example of a 256 QAM radio operating in a 56-MHz channel and supporting 350 Mbps of customer available data across the wireless link. When transmitting Ethernet traffic, this link will be transmitting 350 Mbps of Layer 2 Ethernet data. Assuming that a full 1,518-byte Ethernet frame is being used, (4.1) shows that the link can support a Layer 1 input/output line rate of 354.6 Mbps—slightly higher than the useable over-the-air data rate. This additional amount of input/output data is due to the 20 bytes of introductory information (interframe gap, preamble and frame delimiter) which is not available for customer data.

Now consider the same Ethernet wireless link operating with a minimum 64 byte Ethernet frame size. Equation (4.1) shows that now the 350 Mbps Layer 2 link is operating with a Layer 1 line rate of 459.4 Mbps—significantly higher than the useable over-the-air data rate due to the larger relative size of the 20

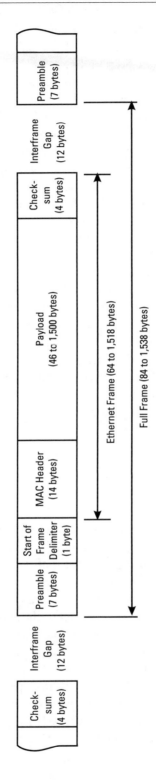

Figure 4.5 Simplified IEEE 802.3 Ethernet Frame.

bytes of overhead against the 64 byte Ethernet frame. Thus, with small Ethernet packet sizes, the wireless link can operate at an input/output line rate of approximately 460 Mbps. However, approximately 110 Mbps of this capacity is allocated to frame overhead that is not available as useable capacity for the transmission of customer data. (Note that strictly speaking, from the discussion here and Figure 4.5, it can be seen that in these examples, even the full 350 Mbps is not available for customer data, since some of the transmitted frame is required for source and destination address and control. However, since this information is required as a part of the Ethernet protocol and since just 18 to 22 bytes are allocated, depending on Ethernet features employed, its effect is generally ignored.)

There are therefore many ways in which the throughput of a high capacity device can be quoted. Table 4.2 summarizes several ways in which the throughput of the example 256 QAM, 56-MHz channel size radio link can be specified. The correct convention is to refer to this as a 350-Mbps wireless link.

Despite all these variations in how to specify data rate, there are system architectures that can be employed to double and even quadruple the true capabilities of a single channel microwave link by using different combinations of channel arrangements. These are now considered further.

4.4.2 Adjacent Channel Transmission

For adjacent channel transmission, two adjoining radio channels are employed together to double the allocate bandwidth available to the link. Given that radio systems are designed to operate in a single channel, with characteristics to provide strong adjacent channel rejection, two can be engineered side by side to allow parallel transmission streams, doubling the capacity of the link. Since both carriers are on the same polarization, this technique is often referred to as adjacent channel copolarized (ACCP) transmission.

Table 4.2
Various Ways in Which the Data Throughput for
an Example 350-Mbps Wireless Link Can Be Quoted

Quoted Throughput	Frame of Reference
Gigabit Ethernet	Protocol of the data connector
740 Mbps	Transmitted aggregate (two-way) over-the-air/gross data rate
370 Mbps	Transmitted FDD over-the-air/gross data rate
350 Mbps	Actual customer available data rate
460 Mbps	Ethernet line rate for smallest Ethernet frame sizes (64 bytes)

4.4.2.1   Typical System

A representation of adjacent channel transmission is depicted in Figure 4.6. Here it is shown how two wireless units can be architected to deliver twice the data throughput, or 700 Mbps for a pair of 56-MHz 256 QAM wireless devices. Note, however, that to double the capacity, duplication of all the radio equipment, except the antenna, is required. Thus, two ODUs and two IDUs are required to support this configuration. Some IDUs allow for dual channel optimization, enabling a single IDU to support a pair of ODUs, making the system hardware more cost-effective and efficient. Care must be taken of the data interface, as the two separate data streams need to be combined into a single input/output signal. For example, link aggregation (LAG) techniques are usually used to combine the two 350-Mbps streams into a single 700-Mbps stream in the example shown in Figure 4.6.

In practice, adjacent channel transmission systems are difficult to implement. The filtering in the combiner network is particularly difficult to realize. Consider the pair of transmit filters. These are required to pass the wanted frequency, but to reject the adjacent channel frequency. This requires very sharp

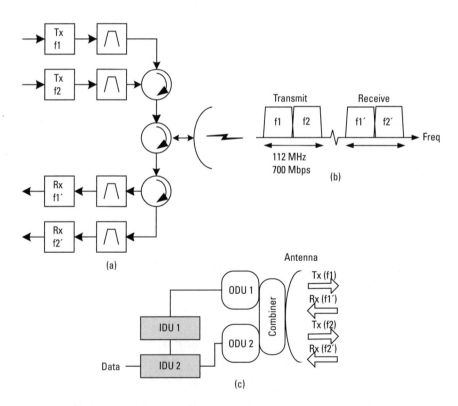

Figure 4.6   Representation of a dual channel 2 × 56 MHz, 700 Mbps data rate device. (a) Block diagram, (b) frequency plan, and (c) typical physical representation.

high-Q filtering with precise tolerances. This can be difficult and costly to design. Other practical considerations are the licensing fees, which are usually twice the cost of the single channel license, the difficulty in obtaining adjacent frequency channels, and the extra loss of the filtering and combining components, which will reduce the system performance from that of a single channel configuration.

4.4.3 Dual-Polarization Transmission

An alternative implementation to double the capacity of a link is to transmit independent channels on dual polarizations. In a single channel microwave link, radio energy travels in waves transmitted in a single horizontal or vertical polarization. It is possible to transmit microwave energy simultaneously on both horizontal and vertical polarizations at the same time, as shown in Figure 4.7. If independent data throughputs are sent on alternative polarizations at the same frequency, twice the data throughout is achieved per frequency channel. This is often referred to as cochannel dual-polarization (CCDP) transmission or sometimes simply as cross-polarization.

4.4.3.1 Typical System

A representation of dual-polarization transmission is shown in Figure 4.8, depicting how 700 Mbps throughput can be achieved for a pair of 56-MHz 256 QAM wireless devices. Note that similar to adjacent channel transmission, two

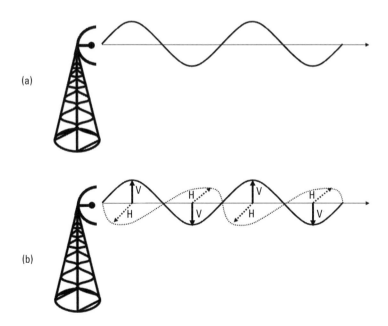

(a)

(b)

Figure 4.7 (a) Single-polarization and (b) dual-polarization transmission.

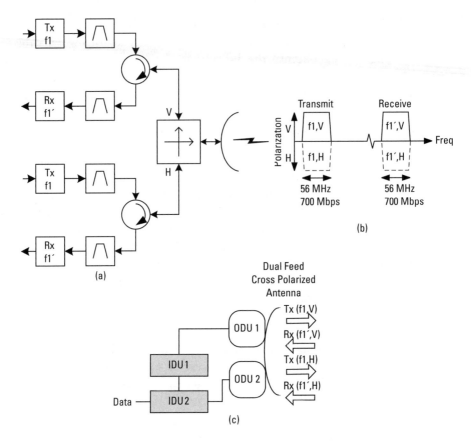

Figure 4.8 Representation of a dual-polarization 56 MHz channel, 700 Mbps data rate device. (a) Block diagram, (b) frequency plan, and (c) typical physical representation.

IDU and ODUs are required, connected to a single antenna. Again, some IDUs allow for a single IDU chassis that supports dual ODUs to simplify the setup and reduce costs and ensure that a single user data channel is available.

A practical dual-polarization system is easier to implement than an adjacent channel system for various reasons. First, the need for a complex filter/combiner is removed. However, in a dual-polarization system, it is replaced with a specially designed cross-polarization antenna. A cross-polarization antenna has two transmission feeds, one for each of the signals from the two ODUs, which it transmits with vertical (V) and horizontal (H) polarizations. The antenna is designed to have minimum cross-polarization discrimination (XPD), a measure of how much one polarized signal will unintentionally leak into the alternative, orthogonal polarization because of imperfect cross-polarization isolation. Second, the difficulty of obtaining two adjacent channels is removed in CCDP transmissions. Finally, licensing is simplified as usually (although not always) no additional license fees are levied when a second cochannel polarization is utilized.

Despite the relative simplicity of implementing a dual-polarization wireless system, there are a number of practical limitations that can degrade the performance of such an architecture. First, launching stable and perfectly horizontal and vertical transmissions is hard to achieve. Antennas and mounts can never be installed for perfect horizontal or vertical transmission, and mounting poles, towers, and even buildings can sway slightly in the wind or twist as the sun heats up and moves across the structure. Second, antennas always have some degree of XPD, whereby the V and H fields are never truly orthogonal, and there will be some V component in the H field and some H component in V field. Third, the channel itself will introduce distortion, especially during atmospheric events such as rain fades. Finally, the receiving antenna is difficult to perfectly align to the transmitter antenna, given the long distance between units and imperfect mounts and installations. Therefore, there are a number of sources of degradation in a cross-polarized link that will reduce system performance over that of copolarized transmissions.

For short-distance links, this overall degradation can be small and there is no need to overcome the effects. Microwave links utilizing just high XPD antennas can operate with acceptable performance. For longer links, however, where these degradations can accumulate and the system margin against such system imperfections is reduced, cancellation techniques are required in order to operate the link with high levels of system availability. Two techniques are available to overcome these limitations: polarization alignment filters and cross-polarization interference cancellation (XPIC).

4.4.3.2 Polarization Alignment Filtering

The systemic errors indicated above can be tuned out by using polarization alignment filters. These allow the V and H polarizations to be set exactly orthogonal to one another (so that there are no V components in the H plane and vice versa), and also to ensure that the orthogonal V and H planes are properly aligned between the two antennas. Polarization alignment filters require manual tuning, adding additional time and care during system install and commissioning, but they do allow all degrading equipment effects to be tuned out, increasing the system performance of the dual-polarization microwave link. Polarization alignment filters, however, do not compensate in any way for any cross-polarization degradation caused by the wireless channel.

4.4.3.3 XPIC

Cross-polarization interference cancellation (XPIC) is a technique that enables all forms of cross-polarization interference to be cancelled. It is able to compensate for any installation or equipment systemic errors and also for any cross-polarization interference caused by fading, dispersion, or other atmospheric effect on the transmission channel between the two wireless devices.

A block diagram of an XPIC system is shown in Figure 4.9. The two separate, independent data streams are transmitted over the same wireless channel using a single, dual-feed antenna with separate V and H polarizations. Despite the orthogonality of the two signals, some interference between the transmissions inevitably occurs due to imperfect system installation, antenna isolation, and channel degradation. XPIC systems function by filtering the individually polarized signals and generating a cancellation signal. This cancellation signal can then be added to the received signal to subtract the cross-polarization interference from the wanted polarization signal, allowing the receiver to successfully regenerate the original desired signal.

The first wireless systems to implement XPIC did so with analog circuitry at the IF frequency. Modern systems, however, implement XPIC digitally, usually within the modem FPGA or ASIC device.

4.4.4 Dual Channel Dual Polarization

In order to achieve gigabit per second and greater speeds, systems can be implemented with a combination of adjacent channel copolarization (ACCP) and co-channel dual-polarization (CCDP) to achieve a fourfold increase in data rate over the standard single channel architecture. This topology is known as dual channel dual polarization (DCDP).

4.4.4.1 Typical System

A representation of dual channel dual polarization transmission is shown in Figure 4.10. Here it can be seen how four wireless units can be architected to deliver four times the data throughput of a single radio, or 1.4 Gbps for four 56-MHz 256 QAM wireless devices. Note that as for previous methods to increase capacity, multiples of radio equipment are required. In this case, four ODUs are required, along with a coupler, to enable a single dual-feed low XPD antenna to be used. IDUs are often optimized to support the four ODUs required for this operation. Difficulty often arises on the data interfaces, as the total data capacity for a DCDP configuration will likely exceed 1 Gbps, meaning that two GbE ports are required to support the full wireless throughput.

As shown earlier, adjacent transmission systems are difficult to implement because of the filtering requirements in the combiner. However, practical systems that use dual channel dual-polarization architectures are commercially available that provide data rates in excess of 1.4 Gbps.

4.4.5 Data Compression

As shown in Section 4.4.1.3, techniques to remove the headers in standard Ethernet frames are an effective way to gain a small amount of improvement in effective data throughputs. To increase data rates even further, additional techniques

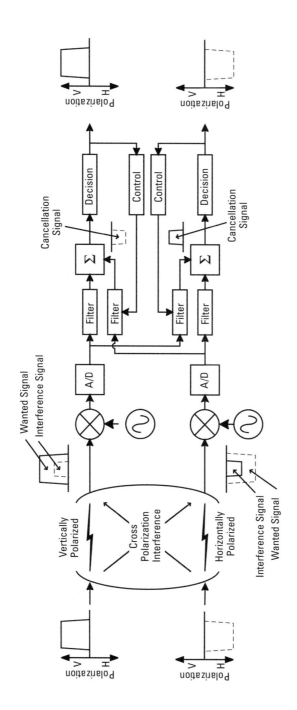

Figure 4.9 XPIC system architecture and operation.

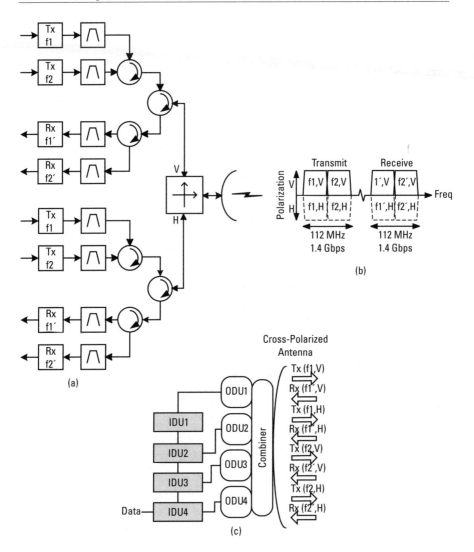

Figure 4.10 Representation of a dual channel dual-polarization 1.4-Gbps data rate device. (a) Block diagram, (b) frequency plan, and (c) typical physical representation.

such as data compression can be employed. Technically, these do not increase the data rate transmitted over-the-air, since this gross bit rate is a function of the channel bandwidth, the number of channels and polarization used, and the operating parameters of the radio. However, data compression does provide the user with more input data capacity to fill the fixed capacity transmitted over the wireless channel. As such, these techniques provide effective throughput increases.

Data compression is often marketed as capacity optimization. Systems are available that employ lossless compression to enables the effective throughput of a high data rate link to be increased up to 2.5 times. One example is allowing a CCDP system to provide up to almost 2 Gbps effective throughput in a single 56-MHz channel. For a DCDP configuration, such a system can provide up to almost 4 Gbps of effective data throughput.

Such compression is strongly dependent on the nature of the data traffic. Compressed Voice over IP (VoIP), Hypertext Markup Language (HTML), and File Transfer Protocol (FTP) traffic is already heavily compressed and cannot be compressed much further. However, uncompressed VoIP can be readily compressed and partially filled TDM traffic frames can be optimized for data throughput. Adding such compression techniques does allow more effective data throughput, but it also adds latency to the data traffic and signal processing complexity to the wireless architecture.

4.4.6 MIMO

Multiple-input multiple-output (MIMO) is the term used for a wireless transmission system that employs multiple antennas at both the transmitter and receiver to improve communication performance. MIMO technology is currently a high profile research area since it offers significant data throughput and link range increases in wireless systems without adding additional bandwidth or transmit power. The more advanced WiFi and WiMAX standards (IEEE 802.11n and IEEE 802.15m respectively) and the LTE 4G cellular standard employ MIMO technology to achieve improved data rates over previous generation systems.

MIMO achieves these performance gains by transmitting separate data streams from different antennas within the same frequency channel. When the multiple, uncorrelated signals arrive at the receiver, sophisticated signal processing techniques are used to detect the different data streams, recombine, and regenerate the original data input. This technique allows higher system spectral efficiency and improved link reliability, which, in turn, allows longer distance transmissions.

For MIMO to be effective, a high degree of multipath reflection is required, for example, off-building in an outdoor environment or off-walls in an indoor application. This scattering is necessary so that each antenna has a different view of the multiple transmitted signals, giving a different distribution of signal amplitudes and phases as signals arrive at each antenna delayed in time from the multiple reflections. For this reason, MIMO operates well for lower frequency PTMP systems, where multipath is prevalent. For PTP microwave links, especially at high data capacities, links are engineered for a minimum of multipath reflections. Line of sight is required, and high directivity antennas are used. For this reason, MIMO is not appropriate for high data rate microwave wireless transmissions.

Despite this, some radio manufacturers have undertaken research on MIMO techniques for high capacity microwave radios, especially in protected diversity applications where an additional radio path is engineered in parallel with the principal path, but usually not operated as it is intended as a standby in case of primary path failure. Such experiments have been shown to yield small improvements in throughput, but generally the increases in system complexity and much harder system installation and commissioning outweigh the advantages. At present, no such commercial high data rate MIMO microwave systems are available.

4.5   The Future?

4.5.1   Higher-Order Modulations

As shown already in this chapter, the current state of the art in microwave radio data throughput is achieved using systems employing 256 QAM modulation. Increasing modulation to 512 QAM or even 1,028 QAM will bring about an increase in capacity. However, there are advantages and disadvantages with such a move.

On the positive side, increasing modulation from 256 QAM to 512 QAM will improve modulation efficiencies from 8 b/s/Hz to 9 b/s/Hz, which will increase data throughput by around 11%. However, this comes with a significant penalty in link performance. For example, the required signal-to-noise ratio (SNR) of the link has to be increased, resulting in a reduction in system sensitivity, which, in turn, will result in shorter links for a given set of performance characteristics. Also, the phase noise and linearity of the microwave transmitter and receiver will need to be increased, adding to the system complexity and probably cost.

At the current time, although modems are available with 512 QAM modulation, no commercial high data rate 512 QAM systems are commercially available.

4.5.2   Channel Bonding

Because of the need for higher data capacities, and the necessity for wide channels to support this, regulators will start allowing channels to be bonded together, forming single 100-MHz or 112-MHz channels, rather than pairs of aggregated 50-MHz or 56-MHz channels. Having these wider channels will remove many of the burdens of ACCP transmission. In particular, the need for a costly and difficult-to-manufacture combiner will be removed. The release of wider channels will enable vendors to build more cost-effective, higher data capacity single channel microwave radio products.

4.5.3 Higher Frequency Bands

Currently, the microwave bands up to 40 GHz are widely used around the world for high data rate communications. In Europe there are other bands at higher frequencies that are beginning to receive attention as spectrum in the lower frequency bands becomes more heavily used. In particular, the 52-GHz band (51.4 to 52.6 GHz) and the 55-GHz band (55.78 to 57.0 GHz) are gaining attention, since they are both relatively unused, and both allow channel sizes up to 56 MHz. Unfortunately, these bands are not available for commercial use in the United States.

Of these, the 52-GHz band is the more attractive since it has better propagation characteristics. As shown in Figure 2.5, transmissions at 50 GHz and above are subject to oxygen absorption in the atmosphere. At 52 GHz, this is still effectively negligible at less than 1 dB/km, but at 55 GHz, atmospheric attenuation raises to 4 dB/km at sea level, meaning that practical link distances are limited.

As shown in the following chapters, there is also significant interest in using the higher millimeter-wave bands at 60 GHz and 70/80 GHz for high capacity wireless communications.

4.6 Summary

In general, the higher the operating frequency, the more spectrum and bandwidth are available and the greater the data rate that the frequency band can support. The microwave bands from 6 to 40 GHz can support full-duplex data rates of several hundred Mbps up to 1 Gbps and higher. The lower frequency bands of less than 6 GHz are widely used for communication purposes, but cannot support such data rates.

There are many systems and applications that operate in the microwave frequency bands. Satellite, radar, traffic monitoring and control, and many other commercial and defense systems utilize these frequencies, but none offers the high-speed data capacity of PTP fixed wireless systems. Such systems are widely deployed for connecting together cell sites, buildings, campuses, and other enterprises. Fixed wireless systems of all speeds have been commercially available for more than 30 years, and in 2009 alone, over 1 million such systems were sold worldwide.

Fixed wireless microwave radios are required to operate in channel sizes up to 56 MHz in Europe and 50 MHz in the United States. Such systems have to be type approved or authorized by the European agencies or FCC, respectively, and licensed before use. Within such channels, systems are commercially available that use high-order 256-QAM modulation to support full-duplex data rates to around 350 Mbps or so. System architectures using parallel links operating in adjacent channels or with dual-polarizations within the same frequency channel, or even both, can be used to double and even quadruple the effective data throughput. These systems can yield full-duplex data rates in excess of 1 Gbps. Techniques such as (XPIC) can be used to overcome system limitations and enable long transmission distances to be achieved.

References

[1] Manning, T., *Microwave Radio Transmission Design Guide*, 2nd ed., Norwood, MA: Artech House, 2009.

[2] FCC, "Code of Federal Regulations, Title 47—Telecommunication, Part 15: Radio Frequency Devices," 2009.

[3] FCC, "Code of Federal Regulations, Title 47—Telecommunication, Part 101: Fixed Microwave Services," 2009.

[4] Six multipart standards under the generic title "ETSI EN 302 217, Fixed Radio Systems; Characteristics and Requirements for Point-to-Point Equipment and Antennas."

[5] 32 multipart standards under the generic title "ETSI EN 301 489, Electromagnetic Compatibility and Radio Spectrum Matters (ERM); Electromagnetic Compatibility (EMC) Standard for Radio Equipment and Services."

[6] EN 60950-1, "Information Technology Equipment—Safety—Part 1: General Requirements."

[7] J. Stuparits, "Backhaul Requirements: Year 2015," Verizon Wireless presentation, 2010.

[8] Morgan Stanley, *The Mobile Internet Report*, 2009.

[9] IEEE 802.3-2008, "IEEE Standard for Information Technology-Specific requirements—Part 3: Carrier Sense Multiple Access with Collision Detection (CMSA/CD) Access Method and Physical Layer Specifications," 2008.

5

Multi-Gigabit 60-GHz Millimeter-Wave Radios

5.1 Introduction

Millimeter-wave communications hold the key to the next phase of the ongoing wireless revolution, in which wireless devices catch up with the fastest wireline speeds. The many gigahertz of unlicensed spectrum in the 60-GHz band is well-suited to short-range multi-gigabit per second transmissions over indoors distances of about 10m, and outdoors over a 500m–1 km range.

The 60 GHz corresponds to an oxygen absorption peak, resulting in a large attenuation of radio waves through the atmosphere (see Figure 2.3). This effect limits propagation distances, but brings the benefits of frequency reuse and minimizing interference. The 60-GHz transmissions require a line of sight between devices, meaning that links need an obstruction-free path between antennas. Antenna gain and directivity increases with higher operational frequencies, so high gain antennas for millimeter-wave communication links can be realized at 60 GHz with footprints much smaller than the lower-frequency microwave bands.

Two distinct applications exist for high data rate 60-GHz wireless: point-to-point outdoor communications, primarily for medium-distance enterprise connectivity, and the potentially much larger market of short-distance ultrahigh data rate wireless networking via 60-GHz wireless local area networks (WLANs). A variety of standards are available or in development to support and enable

60-GHz WLAN applications, including uncompressed HDTV distribution, gigabit Ethernet wireless networking, and very high data rate transfer between consumer electronic devices. Processing techniques traditionally used at lower frequencies are being pushed to higher frequencies to realize low-cost, highly integrated devices that meet the functionality required for these applications.

This chapter focuses on the 60-GHz millimeter-wave band and explores the band's characteristics, uses, and how it enables gigabit per second and higher data rates. Global frequency allocations, rules, and regulations are explored, and the different markets and applications detailed. Analysis is made of the devices used to realize these functions. Finally, some future high capacity 60-GHz directions are given, highlighting the potential for mass market appeal of multigigabit 60-GHz wireless devices.

5.2   Characteristics of the 60-GHz Band

5.2.1   Frequency Bands and Channel Sizes

Many countries have allocated wide bandwidths of spectrum at 60 GHz that permit the transmission of ultrahigh data rates. Most of these bands have only been allocated in the last few years, and there is little harmonization around the world, meaning spectrum allocations differ on a county-by-country basis.

5.2.1.1   United States

In 2001, the FCC allocated 7 GHz of spectrum from 57 GHz to 64 GHz for unlicensed use. The spectrum is not channelized in any way, meaning users are permitted to use the full 7 GHz of bandwidth. This band represents the largest contiguous block of radio spectrum allocated in the United States.

5.2.1.2   Canada

Frequency bands in Canada are regulated by the Spectrum Management and Telecommunications office within Industry Canada (IC). The 60-GHz band has been harmonized with the United States, and so the same 57–64-GHz frequency limits are permitted for unlicensed use.

5.2.1.3   Europe

In Europe, the allocation and usage of frequency bands are managed by CEPT. The frequency allocation in Europe is very different from that available in the United States and Canada.

As of the time of writing, there are only two bands around 60 GHz formally available for high data rate communications in Europe. The first is the 57–59-GHz band, which has been available for many years for analog and digital PTP radios for licensed High Density Fixed Service (HDFS) applications.

In this band, channel sizes are fixed at 50 MHz and 100 MHz, enough for high data rate transmission. However, technical restrictions such as low output power limits and stringent transmitter power spectral density masks, coupled with the challenging propagation characteristics of 60 GHz, have limited practical 57–59-GHz deployments. The second band available at around 60 GHz is the 64–66-GHz frequency allocation, which is also intended for licensed HDFS systems. Four different channel plans are defined for this band [1]. In the first, a 10-MHz guard band is defined at the top and bottom of the band, and the remaining 1.98 GHz of spectrum is divided into 66 channels of 30 MHz each. Users are permitted to aggregate up to 66 together, to effectively allow the full use of the almost 2 GHz of bandwidth available. In the second channel plan, a 50-MHz guard band is defined at the top and bottom of the band, and the remaining 1.9 GHz of spectrum is divided into 38 channels of 50 MHz each. Users are similarly permitted to aggregate any number together to allow full use of the almost 2 GHz of bandwidth available. The third scenario is very similar to the second, except that the lower 50 MHz guard band is replaced with a similar sized thirty-ninth 50-MHz channel, allowing the band to be joined with the 57–64-GHz band discussed next. The final permitted band plan is free use of the band, whereby regulators may allow users unrestricted use of all or any part of the full 2-GHz assignment. Technical rules for the 64–66-GHz band are a lot more accommodating than for the 57–59-GHz band, making this high frequency allocation a lot more attractive for high data rate communications.

CEPT has also released a recommendation for allocating the frequency band from 57 to 64 GHz for unlicensed use in Europe [2], an allocation similar to the United States and Canada frequencies. Originally, a wider 57–66-GHz band was considered, incorporating both the existing 57–59-GHz and 64–66-GHz bands, but the latter was dropped from consideration because of the availability and acceptance of the band in the United Kingdom and a number of other countries. CEPT have recommended two different channel arrangements. The first is an open system bandwidth occupying the whole 7-GHz band, similar to the allocation in the United States and Canada. The second recommendation is for a 50-MHz channel size, whereby 140 channels are defined covering the 57–64-GHz spectrum. Operators are allowed to aggregate up to 50 channels together, limiting the maximum channel bandwidths to 2.5 GHz.

Although most national European regulators follow CEPT guidelines, they are not mandatory and individual country regulators are permitted to modify CEPT's advice to their own situations. This is evident in the 57–64-GHz band, as each European country is currently determining how to adopt the recommendations. In the United Kingdom, for example, OfCom is undergoing public consultation and is likely to allow unlicensed open-band use of the 57–64-GHz band, while retaining the existing 64–66-GHz band for licensed HDFS applications. Other countries that have not yet opened the 64–66-GHz band may well combine this with the 57–64-GHz allocation to permit 9 GHz of spectrum for high data rate transmissions from 57 to 66 GHz.

5.2.1.4 Other Countries

A number of other countries have also opened other 60-GHz bands. In Japan, the Japanese Ministry of Public Management, Home Affairs, Posts and Telecommunications opened an unlicensed 59–66-GHz band in 2000. The band is not channelized, but regulations are such that maximum transmission bandwidths must not exceed 2.5 GHz. Next to this unlicensed 60-GHz band is a wide licensed band that extends from 54.25 to 59 GHz, which is unique only to Japan. The Australian Communications and Media Authority (ACMA) manages the radio spectrum in Australia. In 2005, the ACMA authorized the use of unlicensed devices in the 59.4–62.9-GHz frequency range for low power transmissions. This was later revised to allow outdoor use in the 59–63-GHz band, with indoor use permitted at 57 to 66 GHz. In New Zealand, the Radio Spectrum Management Department of the Ministry of Economic Development has allocated 5 GHz of bandwidth from 59 to 64 GHz for short-range 60-GHz devices.

Other countries that have allocated wide bandwidths of spectrum at 60 GHz for unlicensed use include:

- Korea: 7 GHz of bandwidth from 57 to 64 GHz;
- Brazil: 7 GHz of bandwidth from 57 to 64 GHz;
- China: 5 GHz of bandwidth from 59 to 64 GHz;
- South Africa: 5 GHz of bandwidth from 59 to 64 GHz.

Figure 5.1 shows a summary of the 60-GHz unlicensed band allocations in many of the major international wireless markets. It can be seen that a core band of 59 to 64 GHz is available in most major countries in the world, allowing a common 5 GHz of bandwidth for global high capacity unlicensed wireless use.

5.2.2 Rules and Regulations

In all countries where wideband 60-GHz transmissions are permitted the 60-GHz bands are released for unlicensed use. Europe is the only exception to this with its unique licensed 64–66-GHz band. Although the bands are unlicensed, the use of wireless equipment is still regulated. Wireless equipment needs to comply with local rules and regulations and has to be authorized by national regulatory agencies to confirm compliance before it can be used or offered for sale. However, once authorized, systems can be installed by end users without the need to obtain a license or pay a fee for use of the device and spectrum. Such equipment is often referred to as license-exempt. Generally, this is in contrast

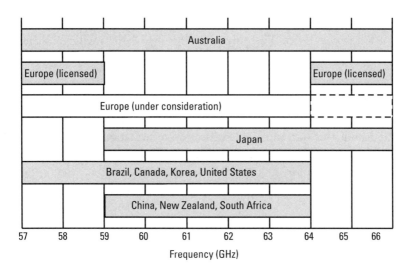

Figure 5.1 Unlicensed 60-GHz frequency allocations in major global markets.

to the lower frequency microwave bands and higher frequency millimeter-wave bands which are almost always licensed.

Similar to the spectrum allocations, technical rules and regulations differ significantly from country to country. The technical specifications in major markets are considered next.

5.2.2.1 United States

The rules and regulations for the 60-GHz band in the United States are governed by FCC Part 15 [3] and in particular section 15.255, which covers the few technical requirements for 57–64-GHz emissions. The FCC allows emissions with an average power density of no more than 9 μW/cm^2 measured at a 3-m distance. This is equivalent to an equivalent isotropic radiated power (EIRP) of 10W or 40 dBm. A peak power density limit of 18 μW/cm^2 measured at a 3-m distance is in place, equivalent to a peak EIRP of 20W or 43 dBm. The FCC also limits the maximum transmit power to 500 mW for any emission with a bandwidth of greater than 100 MHz.

Furthermore, for any indoor transmissions from equipment with more than 0.1 mW of peak output power, the device must transmit an identification signal at least once within a 1-second interval of the signal transmission. This identification signal must contain the following information:

- An FCC identifier, which is hard-coded into the device;
- The manufacturer's serial number, which is also hard-coded into the device;

• A user-defined field of at least 24 bytes of field programmable data relevant to the device. The purpose of this field is to provide information to allow the device operator to be contacted.

The device supplier must publicly specify a method whereby interested parties can obtain sufficient information, at no cost, to enable them to fully detect and decode this transmitter identification information.

Although 60-GHz wireless devices can be freely operated without a license, any apparatus offered for sale in the United States must still have equipment authorization by the FCC, in a similar way as that outlined for microwave devices in Chapter 4.

5.2.2.2 Canada

In Canada, the 60-GHz band has been harmonized with the United States and is managed under Radio Standard Specification RSS-210 [4]. Thus, the frequency limits of 57 to 64 GHz and the rules and regulations regarding the bands within Canada are identical to those in FCC Part 15.

5.2.2.3 Europe

The technical rules for the three European bands are mandated by ETSI. The relevant radio device type approval specification is ETSI EN 302 217 Part 3 for equipment operating in frequency bands where uncoordinated (unlicensed) deployment might be applied [5]. This covers the test methodology and required technical parameters for the 57–59-GHz band, the 64–66-GHz band, and the 57–64-GHz CEPT proposal. Although ETSI breaks out specific requirements for the 57–59-GHz band, it is not considered further in this chapter, as its capabilities are limited and the allocation will soon be incorporated into the wider frequency plan recommended by CEPT. The existing 64–66-GHz allocation and the new 57–64-GHz proposal (ETSI actually considers this to cover 57 to 66 GHz, assuming countries will implement the wider band limits) have requirements with a number of similarities and are considered further in this section.

EN 302 217-3 details the following requirements for 57–66-GHz and 64–66-GHz transmitter equipment, against which compliance is mandatory:

• Spectral efficiency;

• Output power and EIRP;

• Output power tolerance;

• EIRP and/or transmitter spectrum mask;

• Spurious emissions;

• Frequency tolerance;

• Minimum antenna gain.

Although similar categories are defined, specification limits are very different between the wider unlicensed 57–66-GHz band and the narrower licensed 64–66-GHz bands, as both bands are intended to be used for very different applications. The key and unique transmitter requirements for both of these bands are now detailed further.

European 57–66-GHz Unlicensed Band Transmitter Requirements

Consider first the 57–66-GHz band, which is intended for low power, high data rate, unlicensed wireless devices similar to those enabled in the United States and various other parts of the world. Despite approximately equivalent frequency limits, the European device performance requirements are a lot more detailed and generally more stringent than in the United States. For example, a minimal spectral efficiency requirement is placed on 57–66-GHz equipment, as shown in Table 5.1. Spectral efficiency is defined as the ratio between the peak over-the-air bit rate and the radio bandwidth. If the equipment is operating with flexible use of the band, where there are no channels or block arrangements defined, then the equipment is designated as Category 1 equipment and the actual measured transmission bandwidth is used for spectral efficiency calculations. If the equipment uses a single or number of contiguous 50-MHz channels, then it is designated as Category 2 equipment and the bandwidth is the defined channel size. The equipment vendor is free to choose to which category the radio device under test complies.

ETSI specifies a maximum EIRP of 55 dBm (300W) and a maximum output power of 10 dBm (10 mW) for 57–66-GHz equipment. A minimum antenna size of 30 dBi [approximately equivalent to a 3 inch (7.5 cm) parabolic antenna] is also mandated. For equipment operating below the power and EIRP limits, a complex set of relationships between output power, EIRP, and antenna gain are defined. Permitted variations in maximum EIRP and output power with antenna gain are shown in Figure 5.2. Permitted minimum and maximum

Table 5.1
ETSI 57 to 66 GHz Spectral Efficiency Requirements

Equipment spectral efficiency class	1	2	3	4L	4H	5
Modulation	2 state	4 state	8 state	16 state	32 state	64 or 128 state
Minimum spectral efficiency (b/s/Hz)	0.5	1	1.6	2.2	2.8	4

Source: [5].

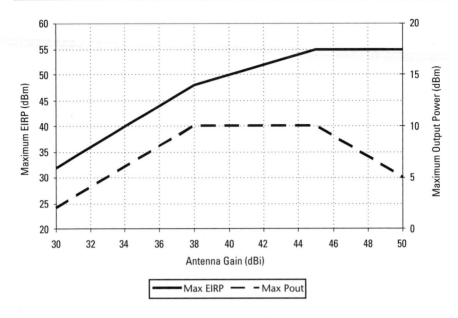

Figure 5.2 ETSI 57–66-GHz requirements for EIRP, output power, and antenna gain. (*After:* [5].)

antenna gain also varies with output power, as shown in Figure 5.3. For equipment operating with the highest allowable output power (10 dBm), only antennas with between 38- and 45-dBi gain are permitted.

The ETSI 57–66-GHz power spectral density emissions mask is shown in Figure 5.4. The mask is not defined as either an EIRP or transmitter mask as it is intended to be equally applicable to equipment with integrated antennas or equipment with a transmitter output connector intended to be connected to an external antenna.

There are several specific considerations to note with this mask in order to make valid measurements. First, the mask is dependent on the occupied bandwidth. For Category 1 equipment, this is the measured transmission bandwidth. For Category 2 equipment it is the defined channel size, which is a multiple of the 50-MHz base channels. Second, the emissions limits are defined as power densities rather than power, meaning that the limit depends on the bandwidth of the emission in addition to its radiated power. Third, the mask is not inclusive of frequency tolerance, which the tester needs to account for. Finally, the maximum 250% limit applies only for aggregate channel sizes of 500 MHz or less. For larger channels, this limit is 150% + 500 MHz.

There are two limitations for 57–66-GHz system out-of-band (OOB) emissions. Although the occupied bandwidth should remain within the specified 57–66-GHz band, some OOB emissions (i.e., those exceeding the 50% abscissa in the spectral density mask) of systems operating close to the band edges may fall outside the band edges. The first OOB emissions requirement is that any

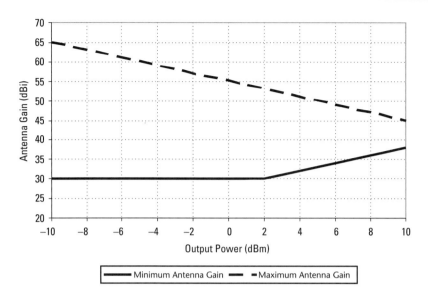

Figure 5.3 ETSI 57–66-GHz requirements for maximum and minimum antenna gain. (*After:* [5].)

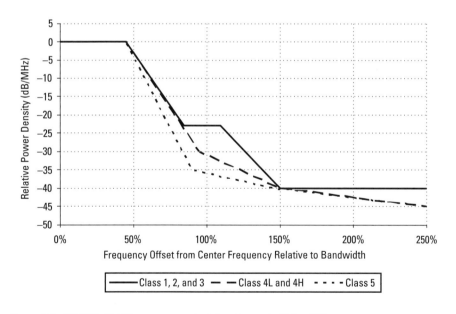

Figure 5.4 ETSI 57–66-GHz power spectral density mask. (*After:* [5].)

of the power spectral density emissions falling outside of the 57–66-GHz band edges shall be limited to a maximum of +10 dBm/MHz. The second OOB requirement is that all other spurious emissions comply to CEPT ERC/Rec 74-01

[6], which is similar for all PTP radio links of any frequency. For equipment operating above about 23 GHz, this requires that all spurious emissions outside the OOB limits defined above be less than −30 dBm. Compliance is required from 30 MHz to the second harmonic of the operating frequency.

European 64–66-GHz Licensed Band Transmitter Requirements

Now consider the alternative 64–66-GHz band permitted in Europe. Here the specification limits are much higher than for the wider 57–66-GHz band, since this band is designed to support longer distance PTP licensed HDFS applications.

The 64–66-GHz spectral efficiency requirements are almost identical to those for the wider 57–66-GHz band. However, Category 1 devices (equipment operating within an undefined channel bandwidth) can only operate with up to level 4 modulation (for example, 4 FSK or QPSK). Category 2 (defined channelization) devices can operate with any modulation. This requirement is intended to encourage channelization for higher density, higher modulation efficiency, HDFS applications.

The output power, EIRP, and antenna gain requirements for 64–66-GHz equipment are as complex as for the wider band 60-GHz equipment, but much higher levels are permitted. A maximum EIRP of +55 dBW (300 kW) is specified, with a maximum output power of 35 dBm (3W) allowed. These are 30 dB and 25 dB higher, respectively, than the 57–66-GHz values. The minimum permitted antenna size remains the same at 30 dBi. Both maximum EIRP and output power vary with antenna gain as shown in Figure 5.5. Note that automatic

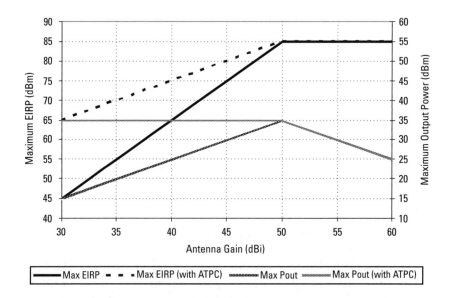

Figure 5.5 ETSI 64–66-GHz requirements for EIRP, output power, and antenna gain. (*After:* [5].)

transmit power control (ATPC) is allowed, although not mandated. For systems with ATPC, both EIRP and output power limits also vary with antenna gain. In addition, a range of antenna gains are permitted, depending on output power, as shown in Figure 5.6. The smallest antenna size can only be used for output powers up to 15 dBm. Above this, minimum antenna size increases. For the highest power systems (35 dBm), only a fixed 50-dBi gain antenna is permitted.

The ETSI 64–66-GHz power spectral density emissions mask is shown in Figure 5.7. Unlike the transmitter masks defined for microwave equipment and the neutral mask for the 57–66-GHz band, the 64–66-GHz requirement is for EIRP emissions and so defines radiated power from the antenna rather than electrical power at the transmitter port. Therefore, for equipment without integrated antennas, compliance is required to a transmitter output power mask derived from this EIRP mask, but reduced by the maximum associated antenna gain. Also, unlike the lower frequency microwave specifications and the wider 57–66-GHz band, the defined emissions spectrum is an absolute mask rather than a relative mask. As such, lower output power 64–66-GHz devices have more margin to the emissions mask than higher power devices. This is contrary to most other mask measurements where the spectrum is measured relative to the output power of the transmitter. The details of interpreting this mask are the same as for the wider 57–66-GHz mask and so are not repeated here. OOB emissions for the 64–66-GHz bands are also exactly the same as for 57–66-GHz bands.

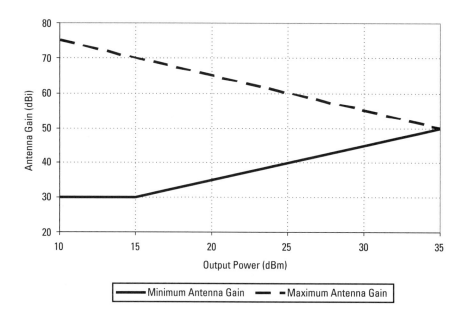

Figure 5.6   ETSI 64–66-GHz requirements for maximum and minimum antenna gain. (*After:* [5].)

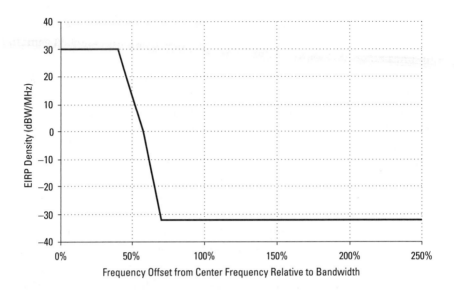

Figure 5.7 ETSI 64–66-GHz power spectral density mask. (*After:* [5].)

European 57–66-GHz and 64 to 66 GHz Receiver Requirements

The ETSI specification EN 302 217-3, which specifies the 60 GHz transmitter characteristics, also defines the receiver requirements for 60-GHz wireless systems. These requirements include:

- Bit error rate (BER) as a function of receiver input signal level (RSL);
- Cochannel interference sensitivity;
- Adjacent channel interference sensitivity;
- CW spurious interference.

Unlike the transmitter requirements, these receiver limits are indicative. Therefore, it is not strictly necessary for equipment to comply with these limits, although properly designed devices should at least meet the given parameters. Limits for only the 64–66-GHz band are given, meaning that there are no requirements, or even guidance, on receiver parameters for the wideband applications. The 64–66-GHz receiver requirements are straightforward and consistent with the lower frequency microwave specifications. As such, they are not considered further here.

European 60-GHz Antenna Requirements

Similar to all other European microwave devices, EN 302 217 Part 4-2 defines the antenna characteristics for 60 GHz systems [7]. No distinction is made between the different 60-GHz subbands, and similar characteristics are required for all equipment. Two classes of antennas are defined, class 2 and 3, each of increasing antenna directivity. Class 3 is further partitioned into a class 3A, which operates with vertical polarization only, and class 3B, which permits dual-polarized vertical and horizontal polarizations. (A class 1 60-GHz antenna for use only in the 57–59-GHz subband is also defined in Part 4-1 of this specification series, but its use is not permitted within the EU R&TTE framework.) Figure 5.8 shows the defined radiation pattern envelope (RPE) requirements for 60-GHz radios for the two single polarization classes of antenna. The ETSI specification also provides guidance that 27 dB of antenna cross-polarization distortion (XPD) should be targeted.

Other European 60-GHz Requirements

Although the 60-GHz subbands have different radio specifications from the microwave bands, the same EMC and safety specifications apply for all these radio types. All 60-GHz equipment needs to carry a CE mark, obtained by formal compliance against the EN 302 217 type approval documents and the same relevant EMC and safety specification, before devices can be sold in Europe.

5.2.2.4 Japan

In Japan, wireless standards are published by the Association of Radio Industries and Businesses (ARIB). Regulations for the 59–66-GHz band were first

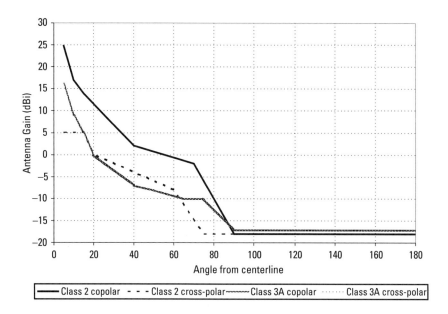

Figure 5.8 ETSI 60-GHz RPE requirements. (*After:* [7].)

set in 2000 and are outlined in STD-T74 [8]. This defines that the maximum transmit power for unlicensed use is limited to 10 dBm, with a maximum allowable antenna gain of 47 dBi. In the adjacent 54.25–59-GHz licensed band, output power is similarly limited to 10 dBm, but much lower antenna gains of only 20 dBi are permitted, limiting licensed applications to just short-distance transmissions.

5.2.2.5 Australia

The Australian Communications and Media Authority (ACMA) defines both indoor and outdoor bands with different power limitations. In the outdoor 59–63-GHz band, the maximum transmitter power is limited to 20 mW and an EIRP of no more than 150W is allowed. In the indoor 57–66-GHz band, a similar transmitter power limit of 20 mW is in place, but EIRP is restricted to a much smaller 20W. Specific requirements for these bands are detailed in [9].

5.2.2.6 New Zealand

In New Zealand, the 59–64-GHz allocation has technical rules equivalent to the United States and Canada. Total peak transmitter power is restricted to no more than 500 mW, and EIRP is limited to 40 dBm average and 43 dBm peak [10].

5.2.2.7 Summary

Table 5.2 summarizes the key technical specifications in the countries considered in this chapter. It can be seen that there are a wide variety of technical rules in place, suggesting that different countries are enabling different types of services. The European 64–66-GHz band, which permits the highest output power and highest EIRP, is clearly targeted towards longer-distance PTP applications. All the other frequency allocations are much wider in bandwidth, suggesting they

Table 5.2
Technical Specifications for Unlicensed 60-GHz Bands in Different Countries

Parameter	United States	Canada	Europe	Europe	Japan	Australia		New Zealand
Indoor/ outdoor	Both	Both	Both	Both	Both	Indoor	Outdoor	Both
Frequency	57–64 GHz	57–64 GHz	64–66 GHz	57–66 GHz	59–66 GHz	57–66 GHz	59–63 GHz	59–64 GHz
Maximum transmitter power	27 dBm for emissions >100 MHz	27 dBm for emissions >100 MHz	35 dBm	10 dBm	10 dBm	13 dBm	13 dBm	27 dBm for emissions >100 MHz
Maximum EIRP	40 dBm average, 43 dBm peak	40 dBm average, 43 dBm peak	85 dBm	55 dBm	57 dBm	43 dBm	51.8 dBm	40 dBm average, 43 dBm peak

are enabling much higher data rate communications. The United States allocation has the smallest EIRP limits, meaning that only low power, short-distance communications can occur. (Interestingly, the U.S. transmitter power limits are relatively high, somewhat inconsistent with the low EIRP values.) Canada and New Zealand have technical rules harmonized with the United States and so also only permit low power, short-distance communications. Other countries (Europe, Japan, and Australia) place relatively low transmitter power limits, but allow larger antennas for longer-distance communications.

5.2.3 Propagation Impediments

As shown in Chapter 2, 60 GHz corresponds to an atmospheric absorption peak due to absorption by oxygen molecules in the air. At sea level, 60-GHz transmissions experience 15 dB/km attenuation (see Figure 2.3), resulting in significant attenuation through the atmosphere. This propagation attenuation places an effective limit on outdoor transmission distances.

When used in an indoor environment, 60-GHz signals are strongly attenuated by solid obstructions, further hampering transmissions. Table 5.3 shows the attenuation through various solid objects commonly found in indoor home and office environments [11]. The same reference also analyzed the effects of humans on 60-GHz propagation. The statistical analysis of four different body types and sizes (two men and two women) yielded a mean attenuation value of 22 dB.

Multipath is an important characteristic for indoor 60-GHz operation. The obstructions listed in Table 5.3 are highly reflective to 60-GHz signals. This phenomenon can be used effectively to aid indoor system design and use, as shown later in this chapter.

Table 5.3
Measured Attenuation at 60 GHz from Various Obstructions

Material Description	Thickness	Attenuation
Simple glass	5 mm	3.5 dB
Double glass	15 mm	4.5 dB
Plywood panels	0.5 cm	6 dB
Metal closet (3-mm galvanized steel plus 5-mm glass)	8 mm	8 dB
Whiteboard (wooden chipboard covered with melamine)	1.5 cm	11.6 dB
Wooden closet (1.5-cm wooden chipboard covered with melamine plus 5-mm glass)	2 cm	13.8 dB
Brick wall with plasterboard on both sides and paint	23 cm	48 dB
Iron door covered with thin layer of plywood panels	8 cm	49 dB

Source: [11].

5.3 60-GHz Market Applications

There are two main applications for high data rate 60-GHz millimeter-wave systems. The first is for high data rate PTP communications, primarily for outdoor wireless systems. This is an established market, with commercial products available from several vendors for this application. The second high data rate 60-GHz market is for high-speed WLANs, applications such as high-speed PC interconnect, wireless HDTV transmissions, and many other applications. This market is still emerging, with few commercial products available yet. However, the 60-GHz WLAN market has the potential to be significantly larger than the outdoor PTP high data rate market going forward.

5.3.1 Point-to-Point Outdoor Connections

Section 4.3 showed how high-speed microwave radios can be used for PTP communications for wireless backhaul applications such as connecting cell sites, mobile base stations, and enterprises. Potentially 60 GHz is also suitable for all these applications. However, in reality, 60 GHz has not gained traction for cellular or mobile backhaul. This is because 60 GHz is a short-range technology and because it operates as an unlicensed frequency band. Telecommunication carriers who own networks and generate significant revenues from these circuits are averse to using unlicensed wireless links where they do not have control over the spectrum and transmission medium and where there is any risk of interference which might interrupt revenue generating services.

The 60-GHz PTP wireless, however, is an excellent solution for enterprise connectivity. Many 60-GHz links are in place connecting and extending LANs between businesses, schools, hospitals, campus buildings, and other enterprises. Since 60 GHz is an unlicensed technology, there is no time or cost associated with applying for and receiving a license, so 60-GHz links can be installed quickly and cost-effectively. Also, because of the very wide bandwidths available, high data capacities can be achieved and can match the standard speeds of 100-Mbps Fast Ethernet and 1-Gbps GbE LANs. The main disadvantage of 60-GHz PTP links for enterprise connections is its limited distance. For distances of up to about 0.5 km, 60 GHz is capable of providing a good solution. For longer distances, system availability will reduce because of the sensitivity to atmospheric effects.

5.3.2 WLAN Connections

With the allocated bandwidth of 5 to 9 GHz of spectrum in many countries, 60 GHz will become the technology of choice for short-range gigabit transmission applications. Due to atmospheric limitations such as line-of-sight requirement and 15 dB/km oxygen absorption, 60-GHz wireless is ideal for short-distance, indoor applications. Also, given that 60 GHz is an unlicensed technology, adoption for consumer applications is greatly enhanced.

A number of 60-GHz applications are currently being actively explored by academic and commercial researchers. These include high-definition multimedia interface (HDMI) cable replacement for uncompressed HD video streaming, mobile distributed computing, wireless docking stations, gigabit Ethernet file transfer, wireless gaming, and more. A summary for the requirements for various WLAN applications is given in Table 5.4.

5.3.2.1 Uncompressed HDTV Streaming

In-home HDTV theatre systems usually consist of a sleek, flat-panel audiovisual (AV) display connected to a variety of HD sources. These might include a cable or satellite set-top box, Blu-ray HD disc player, gaming console, and a broadband Internet connection. An example of such a setup is shown in Figure 5.9. These devices are usually connected by a suite of HDMI cables and switches, which are costly, bulky and unsightly. Although such cables are effective at carrying the high data rates required for HDTV transmissions, there is a market desire to replace the wired interconnections with high data rate wireless, to allow uncompressed, streaming video and sound between multiple sources and viewing devices. High data rate 60-GHz wireless is a leading contender to solve this problem [13].

Very high data rates are required to support uncompressed HD video streaming. The highest quality, commonly available HDTV picture format is currently 1080p. This is so called because it supports 1,080 horizontal lines in a progressive scan (all the horizontal lines are drawn in sequence). With a widescreen aspect ratio of 16:9, 1080p requires a horizontal resolution of 1,920 pixels. This results in a frame resolution of 1,920 × 1,080, or 2 million pixels. With a 60-Hz refresh rate, a three color (RGB: red, green, blue) format and 8-bit resolution per pixel (about 16 million colors), the required data rate for 1080p transmission is approximately 3 Gbps.

Table 5.4
Summary of the Requirements for Various WLAN Applications

Application	Throughput	Latency	Typical Motion
Uncompressed HDTV	3.0 Gbps	Low	Stationary
Wireless projector	2 Gbps	Very low	Stationary
Uncompressed audio	40 Mbps	Low	Mobile
Compressed HDTV	40 Mbps	Moderate	Mobile
Gigabit networks	1 Gbps	High	Mobile
Data transfer	6.0 Gbps	Low	Mobile

Source: [12].

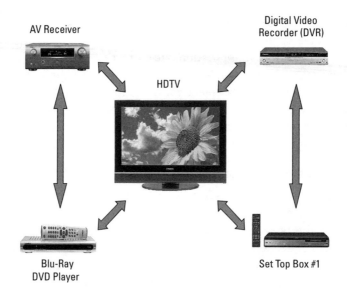

Figure 5.9 Example HD home theater setup. (© 2009 Microwave Journal [13].)

Another common HDTV standard is 1080i, which employs the same number of pixels as 1080p, but the scan is interlaced, meaning that only each second line is refreshed each scan. This allows the system data rate to be halved, but at the expense of some minor picture quality degradation. Other standards such as 720p and 480p are lower-resolution standards. A summary of available HDTV and audio data rates are shown in Table 5.5.

Going forward, the next generation of HDTV will support sharper colors (more bits per color channel) and faster refresh rates. Standards being considered at the moment include 1080p with a 90-Hz refresh rate and 10 bits per RGB

Table 5.5
Required Data Rates for Different HDTV and Audio
Standards

Format	Compression	Data Rate
1080p	None	3.0 Gbps
1080i	None	1.5 Gbps
720p	None	1.4 Gbps
480p	None	500 Mbps
7.1 surround sound	None	40 Mbps
Compressed 1080p	VC-1	36 Mbps
Compressed 1080i	MPEG2	24 Mbps
5.1 surround sound	None	20 Mbps

color, and 1080p with a 60-Hz refresh rate and 12 bits per color. These have resulting data rates of 5.6 Gbps and 4.5 Gbps, respectively.

5.3.2.2 Gigabit Ethernet Wireless Networking

In the wired world, gigabit Ethernet connectivity at 1 Gbps is now displacing 100-Mbps Fast Ethernet as the consumer technology of choice. Most computers nowadays are outfitted with GbE ports, and GbE connectivity is the preferred technology protocol in modern offices. High data rate 60-GHz wireless will allow this wired connectivity to evolve into the wireless domain at gigabit Ethernet speeds.

In the office environment, 60-GHz wireless will allow laptops to connect to the enterprise's LAN and peer computers at speeds equivalent to the wired network. This permits the user mobility and greatly simplifies the office setup. There are already a host of wireline networking technologies that are under consideration for gigabit per second data rates transmission in home networks. For example, G.hn is a collaborative standard being developed by the ITU and promoted by the HomeGrid Forum to support networking over power lines, phone lines, and coaxial cables with data rates up to 1 Gbps. Also, P1901 is a standard developed by an IEEE working group for similar high-speed power line communication standards. Both of these networking protocols can be carried on high-speed 60-GHz wireless networks.

For short distance wireless personal area network (WPAN) applications, high data rate 60-GHz wireless can be used to connect multiple devices for high-speed networking applications, such as rapid backups and remote storage. More traditional, lower data rate applications such as synchronizing data between devices, sending data to a network printer, and connecting multiple peripherals to a device can also be achieved.

5.3.2.3 High-Speed Serial Data Transfer

In contrast to gigabit Ethernet networking, high-speed serial data transfer is for bridging a pair of devices, primarily via a PTP serial interface. Applications include high-speed data transfer and storage of large digital photos and videos or streaming multimedia images from an external hard drive.

Wired data rates and protocols that can be satisfied using 60-GHz wireless include USB 2.0 at theoretical maximum data speeds of 480 Mbps, IEEE 1394 ("Firewire") at 800 Mbps, and SATA (Serial ATA) at 1.5, 3.0, and 6.0 Gbps.

5.4 60-GHz WLAN Standards

It has been shown that high data rate 60-GHz wireless has a significant role to play in WLAN networks for applications such as uncompressed HDTV distribution, gigabit Ethernet wireless networking, and very high data rate transfer between consumer electronic devices. A number of standards are available or in development to support and enable these different 60-GHz applications.

5.4.1   WirelessHD

WirelessHD is a 60-GHz wireless standard specifically architected and optimized for wireless high-definition multimedia connectivity. It is able to connect together consumer electronic (CE) audio and video devices wirelessly. In its first generation implementation, high-speed data rates in excess of 3 Gbps at a 10-m distance can be achieved, permitting uncompressed 1080p AV streaming with a high quality of service (QoS) over a large living room space. WirelessHD's core technology promotes theoretical data rates as high as 25 Gbps, permitting it to scale to higher resolutions, improved color depth, and better operating distance.

The core technology behind WirelessHD was developed by the Berkeley Wireless Research Center around 2002, with a team of academic researchers exploring the high frequency limits of standard complementary metal oxide semiconductor (CMOS) silicon devices. The group developed a process that enabled them to extend operational limits up to 60 GHz and realize wireless devices that can enable high data rate wireless links. In 2004, the core team formed SiBEAM to commercialize their new technology and focus on short-range HDTV distribution. One year later, SiBEAM founded the WirelessHD Consortium to promote the commercial adoption of their technology. The group now includes major CE vendors such as Broadcom Corporation, Intel Corporation, LG Electronics Inc., NEC Corporation, Panasonic Corporation, Philips Electronics, Samsung Electronics Co., Ltd, Sony Corporation, and Toshiba Corporation.

The goals of the WirelessHD Consortium are to develop a standard that meets the following minimum set of requirements:

- Stream uncompressed high-definition video, in addition to audio and data;
- High reliability;
- Low cost;
- Low power solutions for portable devices;
- In-room connectivity (equating to a range of around 10m).

In January 2008, version 1.0 of the WirelessHD specification was released. The specification uses several unique techniques to enable wireless multi-gigabit per second transmission and to support HD uncompressed video streaming. The resulting protocol includes the following attributes:

- High-speed wireless, multi-gigabit technology in the unlicensed 60-GHz band;
- Uncompressed HD video, audio, and data transmission, scalable to future high-definition AV formats;
- Smart antenna technology to overcome line-of-sight constraints;
- Secure communications;
- Device control for simple operation of consumer electronics products;
- Error protection, framing, and timing control techniques for a quality consumer experience;
- Low power options for mobile devices.

The WirelessHD standard is not a public standard and is only available to consortium members. However, the consortium does provide a WirelessHD Specification Overview [14] that is publicly available.

The WirelessHD architecture is based around a number of master and slave devices referred to as coordinators and stations. The setup requires one coordinator (a device that is a sink for audio or video data, such as a display) and any number of stations (devices that can be either sources or sinks of AV media). An example is shown in Figure 5.10.

The high-rate PHY (HRP) is a physical layer that supports multi-gigabit per second throughput at distances of up to 10m. The HRP is a high directivity, unidirectional protocol that uses adaptive antenna technology to steer line-of-sight 60-GHz beams around a room. The low-rate PHY (LRP) is a much lower data rate bidirectional link that provides omnidirectional coverage. All stations support LRP, so it can be used for station to station links. LRP also supports adaptive antenna technology.

In version 1.0 of the WirelessHD standard, the HRP is designed to provide 3-Gbps throughput using orthogonal frequency division multiplex (OFDM)

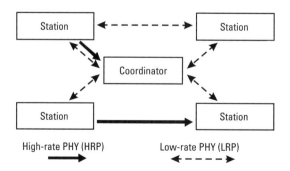

Figure 5.10 Example WirelessHD architecture. (*After:* [14].)

modulation. Multiple data rates are supported by the HRP with various modulations and coding. The LRP is designed to support data rates to 40 Mbps. Key HRP and LRP parameters are given in Table 5.6.

A key attribute of WirelessHD is the adaptive beamforming that is used to ensure connectivity between the station to coordinator and station to station links. Phased-array antennas are used for this purpose. If the direct path between two communicating devices becomes obstructed, the adaptive antenna technology system quickly adapts and finds an alternative path for continuous streaming (for example, bouncing the high data rate content off walls, ceilings, and floors as a person walks across a room and blocks the direct signal path). The WirelessHD protocol consists of two beamforming processes: beam search and beam tracking. Beam search is the process of estimating weighted vectors for the transmitter and receiver antenna arrays that result in the optimum transmission (highest gain or highest SNR) over the 60-GHz channel between the transmitter and receiver. This is used to determine the best alternative path when the preferred transmission path is not available. Beam tracking is the process of tracking the existing transmitter and receiver performance parameters over time due to small perturbations in the wireless channel. This therefore represents continuous adjustment of the antenna to optimized real-time performance. A schematic of the WirelessHD beamforming antenna configuration is shown in Figure 5.11.

Commercially available 60-GHz chips that conform to the WirelessHD standard have been available since the release of the WirelessHD standard in 2008. These are explored in Section 5.6.2.

5.4.2   IEEE 802.15.3c

IEEE 802.15 is the IEEE working group specializing in short range wireless personal area networks (WPANs). This group has developed standards for well-known WPAN technologies such as the first Bluetooth specifications (IEEE

Table 5.6
Key HRP and LRP Parameters for Wireless HD

Parameter	HRP	LRP
Occupied bandwidth	1.76 GHz	92 MHz
Modulation	QPSK, 16 QAM	BPSK
OFDM format	512 subcarriers	128 subcarriers
Forward error correction	Reed Solomon	Convolution coding
Frequency bands	Channel 1: 57.24 to 59.40 GHz Channel 2: 59.40 to 61.56 GHz Channel 3: 62.56 to 63.72 GHz Channel 4: 63.72 to 65.88 GHz	Same as HRP, but with three LPR channels within each HRP channel

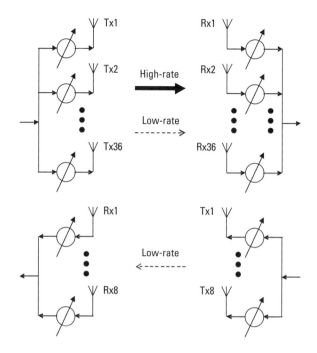

Figure 5.11 WirelessHD antenna array configuration. (*After:* [14].)

802.15.1), ultrawideband (UWB) (IEEE 802.15.3), and the ZigBee standard (IEEE 802.15.4).

The IEEE 802.15.3 Task Group 3c (TG3c) was formed in March 2005 with the remit to develop a millimeter-wave PHY layer for the existing WPAN standard 802.15.3-2003. In particular, the group identified three key attributes that were expected from the development:

• Unlicensed operation in the 60-GHz band;

• Very high data rates, greater than 1Gbps;

• Support for multimedia.

After almost 5 years of consideration, the IEEE announced the formal release of IEEE 802.15.3c-2009 in October 2009 [15]. The standard defines protocols for 60-GHz wireless devices that can support multi-gigabit throughput for CE devices while ensuring coexistence with legacy systems. The standard defines different modes that can enable data rates up to 5 Gbps. In addition, a beamforming protocol is defined to improve the range of communication devices.

IEEE 802.15.3c is different from WirelessHD in that it is a standard for general wireless network connectivity, rather than being a specific standard for a specialized application. As such, IEEE 802.15.3c covers a much wider range of applications, bringing it potentially more universal appeal. Furthermore, IEEE 802.15.3c is an extension an existing, released standard.

As of the time of writing, no commercial 60-GHz IEEE 802.15.3c devices are available.

5.4.3   ECMA-387/ISO/IEC 13156

In December 2008, Ecma International (formerly the European Computer Manufacturers Association), a European developer of standards for information and communications technology and consumer electronics published standard ECMA-387 [16]. Developed in collaboration with Phillips, Intel, Samsung, Panasonic, and others, ECMA-387 provides support for full 60-GHz multi-gigabit per second networking, including HD streaming video applications, fast file download support, wireless docking, and short-distance point-to-multipoint "sync and go" applications for handheld devices. EMCA-387 was submitted for fast-track consideration as the dominant European standard for short range wireless ultrahigh data rate networking. In November 2009, the International Organization for Standardization/International Electrotechnical Commission (ISO/IEC) published the standard as ISO/IEC 13156 [17].

ECMA-387 is unique from the other 60-GHz standards considered so far in that it allows for three different classes of devices, each with different physical implementations that can coexist and interoperate with each other. Thus, it offers a heterogeneous network solution for various devices, each with varying system performance specifications, which allow designers to trade off performance with power, cost, and ease of implementation. A typical network featuring all three devices is shown in Figure 5.12. The three defined devices are as follows:

- Type A devices offers video streaming and WPAN applications over 10-m line of sight (LOS) and nonline of sight (NLOS) multipath environments using high-gain adaptive array antennas. A variety of data rates from 397 Mbps to 6.350 Gbps are supported.

- Type B devices offer video and data PTP transfer over 1–3m using LOS links with fixed antennas. QPSK modulations allow data rates to 1.6 Gbps and optionally to 3.2 Gbps.

- Type C devices support data only applications over PTP LOS links at less than a 1-m range with fixed antennas and no QoS guarantees. Basic OOK modulations support data rates to 1.6 Gbps, with an option for a 3.2-Gbps operation.

The defined ECMA-387 frequency band extends from 57 to 66 GHz, with four 2.16-GHz channels. As shown in Figure 5.13, from one to all four of these

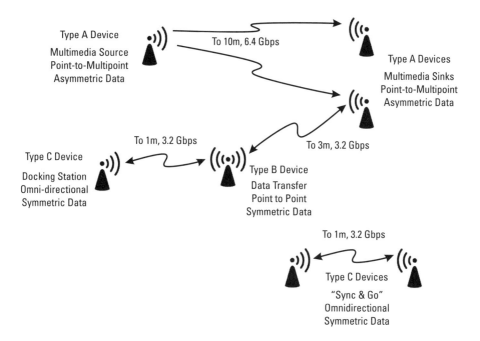

Figure 5.12 Example of an ECMA-387 heterogeneous network.

subchannels can be aggregated together to permit channel sizes up to 8.64 GHz. Via 4-channel bonding, ECMA-387 permits Type A devices to operate at over 25-Gbps data throughput.

Although no commercial chipsets or systems are yet available, a prototype ECMA-387 compliant system showing multi-gigabit per second file transfer and 1080i uncompressed HD video streaming has been demonstrated. EMCA members including the Georgia Electronic Design Center, ETRI (Electronics and Technology Research Institute, Korea), and Philips demonstrated such a system in late 2008. Details of the system setup and detailed performance are not publicly available. However, the 60-GHz RF devices from GEDC have been documented and are detailed in Section 5.6.3.

5.4.4 IEEE 802.11ad

IEEE 802.11 is the IEEE working group specializing in wireless LANs. Over a period of more than a decade, this group has developed a number of standards under the 802.11 WiFi name. These have included commercially successful IEEE 802.11b (2.4 GHz, 11 Mbps) in 1999, 802.11g (2.4 GHz, 54 Mbps) in 2003, and the most recent 802.11n (2.4/5.8 GHz, 600 Mbps) in 2009.

The IEEE 802.11 Task Group ad (TGad) was formed in December 2008 to extend the 802.11 standard to enable operation in the 60-GHz frequency

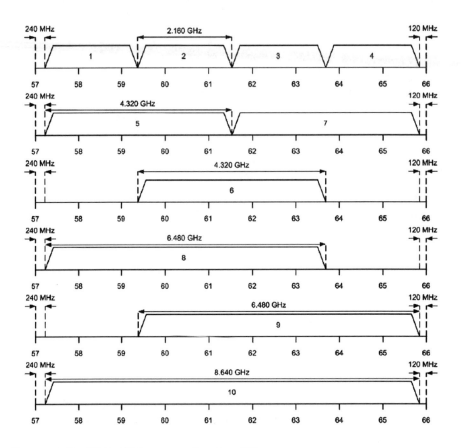

Figure 5.13 The ECMA-387 frequency allocations [16].

band to yield very high data throughput capacities. Specifically, the following four objectives were defined for the group:

- Enable a maximum throughput of at least 1 Gbps;
- Enable fast session transfers between devices;
- Maintain the 802.11 user experience;
- Provide mechanisms that enable coexistence with other systems in the band, including IEEE 802.15.3c systems.

Although not scheduled for release until 2012, IEEE 802.11ad is generating much interest. Since its core technology is based on the already proven and commercially successful 802.11 standard, a solid network architecture is envisioned, which is backwards compatible to the existing 802.11 infrastructure. It is also envisioned that future 802.11 devices will be available which allow operation in the three unlicensed bands of 2.4 GHz, 5.8 GHz, and 60 GHz, enabling ultrafast sessions to exist between devices and networks. The 802.11ad amendment will specify a mechanism for multiband devices, and an additional goal is to strive for a level of coexistence with other 60-GHz wireless standards, particularly IEEE 802.15.3c if feasible.

5.4.5   WiGig

The Wireless Gigabit Alliance (WGA, but more widely known as WiGig) was created in May 2009 with a similar goal of promoting the adoption of multi-gigabit per second wireless communications over the unlicensed 60-GHz spectrum. Like many of the other 60-GHz standards, WiGig is not limited to wireless HDTV distribution, but is designed to be applicable to all wireless networking, from PC interconnects to cable replacement, as well as HDMI cable replacements for wireless HDTV distribution. In addition, WiGig is based on an extension of the IEEE 802.11 WiFi specification, which is already a solid standard and a global commercial success for consumer electronics, handheld devices, and personal computers.

The core of WiGig's technology evolved from Wilocity, a company founded in 2007 by a team of ex-Intel executives and engineers who worked together on Intel's successful WiFi chips. In 2009, Wilocity created the WGA to establish a unified specification for 60-GHz wireless technology. The WGA has a wide range of members from the communications, networking, and computing industry segments, including Atheros Communications, Inc., Broadcom Corporation, Dell, Inc., Intel Corporation, LG Electronics Inc., Marvell International Ltd., MediaTek Inc., Microsoft Corporation, NEC Corporation, Nokia Corporation, NVIDIA, Panasonic Corporation, Samsung Electronics Co., and Toshiba Corporation.

In December 2009, the WGA released version 1.0 of the WiGig specification, which includes the following key elements:

- Supports data transmission rates up to 7 Gbps in the unlicensed 60-GHz bands;
- Supplements and extends the 802.11 medium access control (MAC) layer;
- Backwards compatible with the IEEE 802.11 standard;
- Protocol adaptation layers to support specific system interfaces including data buses for PC peripherals and display interfaces for HDTVs, monitors, and projectors;
- Support for beamforming, enabling robust communication at distances beyond 10m;

• Advanced security and power management for WiGig devices.

The WiGig specification is not publicly available. As of the time of this writing, no commercial WiGig devices are available.

5.4.6 Summary

There are a number of international standards underway or in place to support and enable a variety of different 60-GHz applications. Table 5.7 summarizes these standards, their release date, and the targeted applications. It can be seen that WirelessHD has a head start with devices already available in the marketplace for streaming HDTV applications. Four other standards are competing for the wider WPAN and high data rate networking market, in addition to local HDTV distribution. Three of these four standards are released, but none of them yet supports commercial devices at the time of this writing.

5.5 Technology Choices for High Data Rate 60-GHz Devices

For millimeter-wave operation, III-V technologies are usually the preferred choice for fabricating high frequency devices and components. Materials such as gallium arsenide (GaAs) and indium phosphide (InP) are usually chosen over silicon (Si), the preferred solution for lower frequency devices, because of their higher electron mobility (allowing faster devices to be realized), higher breakdown voltages (allowing higher power level operation), and lower noise. For this reason, GaAs dominates most high frequency applications, with InP offering the best performance for low-noise, high-frequency circuits. Gallium nitride (GaN) is starting to feature as an alternative for high-power, high-frequency devices.

Despite these performance advantages, III-V devices are expensive to fabricate and have relatively low manufacturing yields, resulting in limited integration possibilities. For these reasons, silicon technology is being widely considered for future mass-market 60-GHz applications. Silicon has already displaced GaAs

Table 5.7
Status of the Various 60-GHz WLAN Standards

Standard	Released	Applications	Commercial Units
WirelessHD	January 2008	HDTV streaming	Yes
ECMA-387	December 2008	WPAN, networking, HDTV streaming	No
IEEE 802.15.3	December 2009	WPAN, networking, HDTV streaming	No
WiGig	December 2009	WPAN, networking, HDTV streaming	No
IEEE 802.11ad	2012	WPAN, networking, HDTV streaming	No

and other technologies for RF applications in lower frequency microwave applications. Silicon is cheap, plentiful, and relatively easy to fabricate. It also lends itself to high levels of integration, allowing both digital and analog circuitry to coexist on the same chip.

Complimentary metal-oxide semiconductor (CMOS) is a proven silicon technology for low frequency price-sensitive consumer applications, but its technical performance attributes limit its high-frequency performance. Fortunately, as processing techniques advance and smaller transistor geometries are achieved, the frequency performance improves. Measurements on 90-nm CMOS transistors at the Berkeley Wireless Research Center show an f_T exceeding 100 GHz and an achievable f_{max} over 200 GHz [18]. (f_T and f_{max} are transistor figures of merit, representing the unity current gain and the maximum oscillation frequency of the device, respectively.) Industry is already moving towards smaller geometries of 65 nm, 45 nm, and even 32 nm, but these smaller geometry processes come with increasingly higher processing costs, which limit their use economically to extremely high-volume markets such as microprocessors and other mass-produced devices.

For this reason, there is also serious consideration of silicon variants such as silicon germanium (SiGe) bipolar CMOS (BiCMOS) for high-speed 60-GHz devices. BiCMOS can realize performances similar to CMOS, but uses an older lithographic process that is cheaper to manufacture, even though the base SiGe material is more expensive than bulk silicon.

Given the potential mass market appeal of 60-GHz solutions and the resulting need for low cost devices, packaging of the millimeter-wave components will become an important consideration. In many consumer applications, packaging dominates the bill of materials and exceeds the silicon die costs. For this reason, millimeter-wave packaging will likely be an ongoing area of research, trading off performance limitations of 60-GHz transitions with low cost packages.

5.6 High Data Rate 60-GHz Systems

5.6.1 Outdoor Point to Point

There are a number of systems available for 60-GHz PTP applications. Figure 5.14 shows one such system which provides full duplex data rates up to 1 Gbps in the unlicensed 60-GHz band. This radio product is configured as an all outdoor ODU configuration, whereby the data interface, modem, signal processing, and RF modules are housed in a single outdoor chassis. The radio operates FDD, with transmit and receive frequencies centered on 58.1 and 62.9 GHz, and uses FSK modulation to transmit 1-Gbps of data in a 1.4-GHz occupied bandwidth. An integrated 10 in (25 cm) directional antenna provides 40 dBi of gain with 1.4° beamwidth, which enables the device to deliver a full 1 Gbps of

Figure 5.14 Photograph of a 60-GHz 1-Gbps outdoor PTP radio unit. (*Source:* BridgeWave Communications, 2010. Reprinted with permission.)

data at distances of up to 1 km (0.7 mile) under optimum propagation conditions. The unit is also available with an external 46 dBi, 0.6° beamwidth, 2 ft (60 cm) antenna, for longer-distance transmissions.

5.6.2 WirelessHD

At the time of this writing, only one vendor has commercial WirelessHD devices available. SiBEAM has released two generations of network processor and 60-GHz RF chips to support the WirelessHD protocol. Built using 90-nm CMOS, SiBEAM's devices can provide up to 4-Gbps data transmission in the 60-GHz bands [19]. These low cost silicon chips form the heart of SiBEAM's OmniLink 60 WirelessHD product. Having a silicon solution rather than using devices based on the much more expensive GaAs or InP technologies helps towards the all-important goal of reducing system cost to a price point competitive with HDMI cables.

A block diagram and die photograph of the SiBEAM CMOS millimeter-wave chip are shown in Figure 5.15. The 60-GHz chip architecture contains the complete transmitter and receiver, including low noise amplifiers, power amplifiers, mixers, and lower-frequency IF and baseband amplifiers. The chip also includes 32 integrated 60-GHz antennas that implement the beamforming functions. This complete 60 GHz millimeter-wave chip represents state-of-the-art RF integration, parallelism, and processing for high-volume consumer wireless communications products.

SiBEAM has also developed a network processor chip that complements this 60 GHz RF chip. This chip contains all the digital baseband, mixed-signal components (analog-to-digital converters, digital-to-analog converters, and

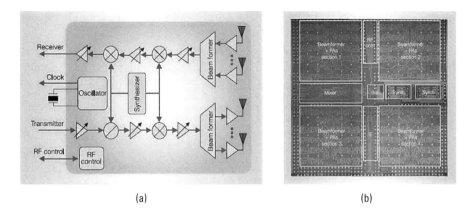

(a)   (b)

Figure 5.15 (a) Radio architecture and (b) die photograph of SiBEAM's integrated 60-GHz chip. (Copyright 2008 IEEE [19].)

phase locked loops), audio and video interfaces, and the embedded microprocessor. The chip also contains the digital OFDM modulator that encodes and decodes transmissions at data rates up to 4 Gbps. The network processor chip controls several functions in the radio chip, including the beamforming settings for the adaptive antennas.

Figure 5.16 shows the SiBEAM network processing and RF chipsets placement in a typical HD scenario connecting together an AV source (e.g., Blu-ray DVD player or HD set-top box) and sink (the HD display). The two chips provide almost full integration of all the circuitry from input AV signals to the RF antenna. Although not shown, only minimal additional circuitry is required to

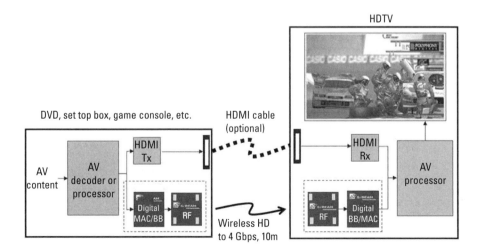

Figure 5.16 Placement of SiBEAM's WirelessHD chipsets in a typical HDMI-replacement audiovisual system. (Copyright 2008 IEEE [19].)

support these two chips, such as voltage supply and regulation, and a frequency reference crystal.

A photograph of a full WirelessHD reference design was shown earlier as Figure 3.3. Vizio, Panasonic, and Toshiba all demonstrated HDTV products incorporating 60-GHz WirelessHD technology at the 2010 International Consumer Electronics Show (CES 2010).

5.6.3   Wireless LAN/PAN

No commercial wireless chipsets are available yet for any of the 60-GHz WLAN/WPAN networking standards discussed in Section 5.4. However, given the recent release of several of these specifications, there is much effort to develop circuits and bring to the market cost-effective CMOS 60-GHz ICs that meet the IEEE 802.15.3c, ECMA-387, WiGig and future IEEE 802.11ad specifications.

Many groups are actively researching this area. Quite a number of 60-GHz chip transceivers have been developed and published in the scientific literature for 60-GHz WLAN/WPAN applications. These have been on SiGe BiCMOS and Si CMOS platforms. A good overview of the state of the art and the relative merits and demerits of different processes and technologies is given in [18]. A few groups have developed fully integrated 60-GHz wireless systems on silicon, including not only the RF transceivers but also the baseband signal processing. The work of two groups developing state-of-the-art, whole-system integrated WLAN/WPAN solutions are further reviewed here.

The Georgia Electronic Design Center (GEDC) recently published an array of silicon CMOS solutions for 60-GHz wireless networking [20]. The group has developed a range of 60-GHz CMOS circuit blocks, all fabricated on 90-nm silicon. These include:

- Two-stage, single-ended low noise amplifier with 6.5-dB noise figure;
- Four-stage power amplifier with a 20-dB small signal gain from 60–65 GHz (the device achieves a P1dB of greater than 8 dBm and a 12-dBm saturated output power);
- Voltage controlled oscillator with a phase noise of −85 dBc/Hz at a 1-MHz offset over a frequency range of 50.7–56.7 GHz;
- Embedded mixed signal processor including mixed signal and OOK, BPSK, and QPSK modulators capable of 1.728- and 3.456-Gbps data throughput.

GEDC have fabricated several complete radio solutions using these building blocks. All are designed for wideband operation over the whole 57–66-GHz band. Operation is targeted towards the ECMA-387 specification, with a channel size of 2.16 GHz at center frequencies of 58.32 GHz, 60.48 GHz, 62.64 GHz, and 64.80 GHz. In addition to the 60-GHz RF circuitry, the designs include on-chip mixed signal processing and up to 200,000 gates. All the 60-GHz CMOS single-chip radios are externally controlled by a standard SPI interface and use a single, low-cost external crystal.

Figure 5.17 shows the architecture for a single chip TDD radio developed by GEDC. The fabricated 60-GHz single chip radio die photograph is also shown. The whole chip occupies just 2.5 × 2.5 mm and achieves up to 3.546-Gbps data throughput. Over the full operational band of 58.32 to 64.8 GHz, the chip achieves a P1dB of greater than 3 dBm and receiver noise figure of 9 dB. The chip operates from dual 1.0- and 1.2-V rails and consumes less than 200-mW power. A comparable FDD chip was also developed that included additional system functionality such as Reed-Solomon FEC. This chip was also fabricated on 90-nm CMOS and occupies a mere 3.7 × 2.4 mm.

To further increase the performance and functionality, GEDC has also developed a 60-GHz phased-array receiver architecture. Although the chipset does not incorporate the actual active antennas as per the WirelessHD devices, it does allow four antennas to be driven from the single chip. Figure 5.18 shows a block diagram of the quad-antenna receiver structure, including phase shifting elements with output amplitude variations of less than ±0.8 dB across a full 360° output phase shift. Die photographs of the receiver with a matching 60-GHz transmitter fabricated on 90-nm CMOS are also shown. Each chip measures 7 × 2.5 mm.

Researchers at the Berkeley Wireless Research Center (BWRC) at the University of California, Berkeley, have also developed a complete 60-GHz wireless system using a similar 90-nm silicon CMOS process [21]. Their architecture is shown in Figure 5.19, along with a die photograph of the complete chip.

The BWRC system has been designed for TDD operation and incorporates both 60-GHz millimeter-wave and mixed-signal baseband circuitry. Operation is targeted at 5–10-Gbps data throughput utilizing QPSK modulation. In the baseband section, the design incorporates a 5-tap decision feedback equalizer (DFE) to compensate for signal interference over the wireless path. The 60-GHz transmitter achieves an output power of 10.6 dBm, and the whole design operates from a single 1.2-V supply and consumes 170-mW power in transmit mode and 138 mW in receive mode. The overall die size is 2.75 × 2.5 mm.

Measurements of the BWRC chip using external 25-dBi horn antennas showed a measured bit error rate of better than 10^{-11} at a 4-Gbps data throughput at a 1-m distance. Improvements in the chip layout and circuitry, particularly the IQ gain and phase mismatch, are expected to improve this performance and distance.

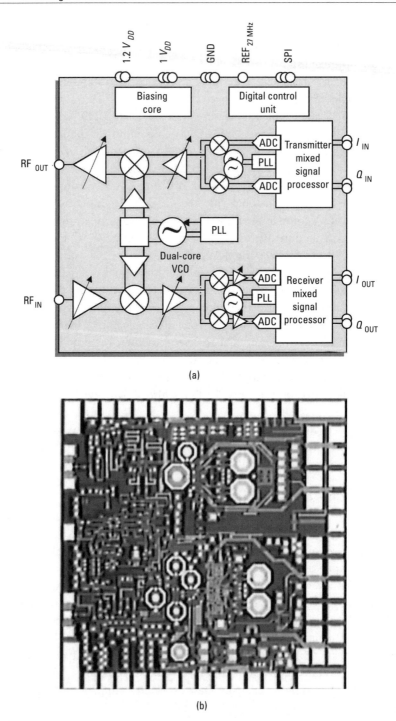

(a)

(b)

Figure 5.17 (a) GEDC architecture for 60-GHz single-chip TDD radio, and (b) die photograph of chip fabricated on 90-nm CMOS. (Copyright 2009 IEEE [20].)

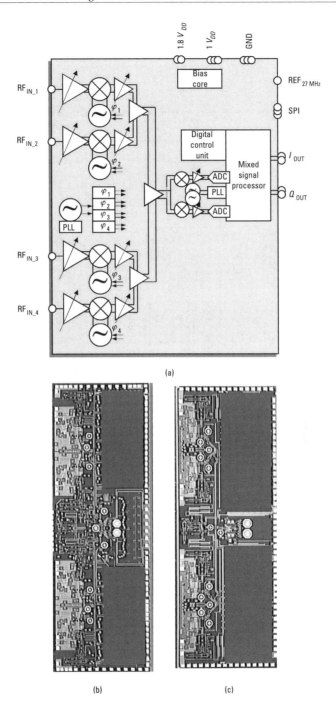

(a)

(b)   (c)

Figure 5.18 (a) GEDC architecture for 60-GHz single-chip phased-array receiver, (b) die photographs of chip fabricated on 90-nm CMOS, and (c) matching 60-GHz transmitter. (Copyright 2009 IEEE [20].)

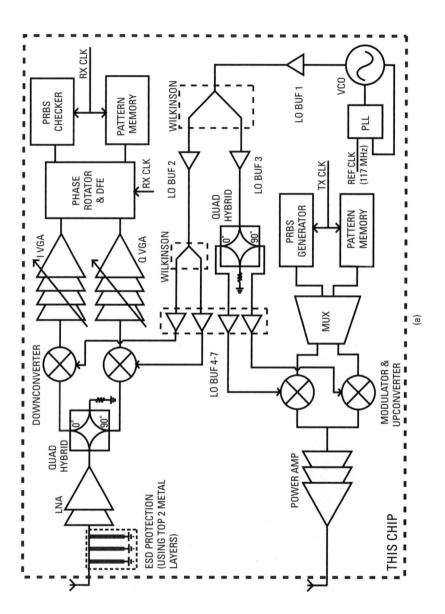

Figure 5.19 (a) BWRC architecture for 60-GHz single-chip radio, and (b) die photograph of chip fabricated on 90-nm CMOS. (Copyright 2009 IEEE [21].)

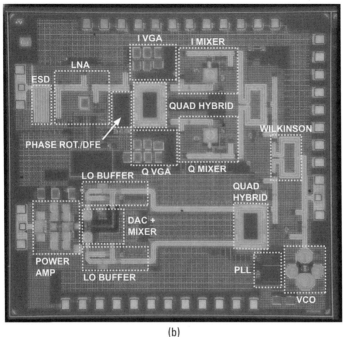

(b)

Figure 5.19 continued

5.7 The Future?

5.7.1 Frequency, Channel, and Specification Harmonization

Perhaps the first advance we will see in the 60-GHz band is closer associations of the various national 60-GHz bands. As shown in Figure 5.1, currently Europe does not formally recognize the wide channels available in many other major markets. With this currently under consideration, and there being CEPT and ETSI recommendations and rules in place for the band, European regulators will quickly see the benefits of harmonizing with the rest of the world and opening the wide channel bandwidths to permit very high data rate, short-distance communications. Once the European 57–64-GHz channel (and perhaps 57–66 GHz) is adopted, there will be a common 5-GHz overlap of spectrum from 59–64 GHz that is available in most of the main consumer markets in the world.

After this, we will see changes in specifications to close the gaps between regional requirements. As shown in Table 5.2, significant specification differences currently exist. For example, the maximum output power allowed for unlicensed wideband transmissions in North America is 27 dBm (500 mW). However, the limit is significantly smaller in Europe at 10 dBm (10 mW). Similarly wide disparities exist in other parameters. The EIRP limit for North America is only 40 dBm, but is much larger at 55 dBm in Europe. Any size antennas (within the above limits) are permitted in North America, but a minimum antenna size of 30 dBi is required in Europe. These current wide variances will hinder device manufactures as different specification units will be required in different regions of the world.

5.7.2 Consolidation of 60-GHz WLAN/WPAN Standards

There are currently five high profile standards released or in development for 60-GHz wireless high data rate wireless communications: WirelessHD, IEEE 802.15.3c, IEEE 802.11ad, ECMA-387, and WiGig. All focus on the similar goals of enabling gigabit per second and higher communications over distances of up to 10m in the unlicensed 60-GHz bands. Clearly there will be winners and losers and likely some consolidation in these standards.

The standardization process is highly political and controversial, with many companies using the process to further their own preferences and slow down (and even shut out) unfavored solutions. This is seen regularly at standards meetings where parties deliberately bring in large contingents to vote en masse and approve or block resolutions in line with the group's beliefs.

Several of the 60-GHz wireless standards have high-profile supporters and wide industry backing. For example, WirelessHD has significant traction with consumer electronics companies, WiGig with the PC industry, and IEEE 802.11ad with the wireless networking community. It is likely that the industry will quickly see a convergence and probably consolidation between WiGig and IEEE 802.11ad, given common goals and overlapping industry backers. The two will likely be incorporated into a future IEEE 802.11 standard, preserving the dominance of this protocol in the wireless LAN world. Also, WirelessHD will likely continue and dominate the more focused high data rate streaming HDTV consumer market (although challenged by lower-frequency 5.8-GHz competitors offering much lower data rate compressed streaming HD solutions). The IEEE 802.15.3c and ECMA-387 standards will have difficulties gaining mass recognition, the former due to its core technology being based on a networking protocol with limited commercial success, and the latter due to its backing by only a limited number of larger sponsors.

5.8 Summary

60-GHz wireless systems offer the ability to transmit extremely high data rates over short distances. There is much interest in this technology as the basis for future wireless networking standards which will offer wireless speeds matching those of available wireline standards. There is potential for large mass market appeal for such devices and systems.

Currently the 60-GHz band is allocated and managed very differently in various countries around the world. Regulators are responding to the need for common frequency bands by adjusting limits to allow harmonization with other countries. Europe in particular is adding significant bandwidth to allow overlap with other nations' 60-GHz frequency plans. However, vast differences in technical specifications and requirements remain, which may be an impediment to mass adoption of higher performing equipment.

Two market applications exist for high data rate 60-GHz wireless: PTP enterprise communications, extending LANs between buildings and across campuses, and short-distance ultrahigh data rate wireless LAN for short-range wireless networking. This latter application has significant market potential, and five international standards are available or in development to support and enable applications such as uncompressed HDTV distribution, gigabit Ethernet wireless networking, and very high data rate transfer between consumer electronic devices.

Traditionally, GaAs and InP materials have been used for fabricating high-frequency devices. Recent advances in Si CMOS technology have enabled researchers to realize low-cost, high-performance 60-GHz devices. The closely related SiGe BiCMOS process can also be used to realize high-speed 60 GHz devices. A number of systems and devices that use these technologies are available. Transmission speeds up to around 4 Gbps have been demonstrated, verifying the potential for commercially realizable, multi-gigabit transmission and mass-market appeal of the 60-GHz millimeter-wave band.

References

[1] ECC Recommendation (05)02, "Use of the 64–66 GHz Frequency Band for Fixed Service," 2009.

[2] ECC Recommendation (09)01, "Use of the 57–64 GHz Frequency Band for Point-to-Point Fixed Wireless Systems," 2009.

[3] FCC, "Code of Federal Regulations, Title 47—Telecommunication, Part 15: Radio Frequency Devices," 2009.

[4] Industry Canada Spectrum Management and Telecommunications Radio Standard Specification RSS-210, "Low-Power Licensed-Exempt Radio Communication Devices (All Frequency Bands): Category 1 Equipment," Issue 7, 2007.

[5] ETSI EN 302 217-3, "Fixed Radio Systems; Characteristics and Requirements for Point-to-Point Equipment and Antennas; Part 3: Equipment Operating in Frequency Bands Where Both Frequency Coordinated or Uncoordinated Deployment Might Be Applied; Harmonized EN Covering the Essential Requirements of Article 3.2 of the R&TTE Directive," version 1.3.1, 2009.

[6] CEPT/ERC/Recommendation 74-01E, "Unwanted Emissions in the Spurious Domain," 2005.

[7] ETSI EN 302 217-4-2, "Fixed Radio Systems; Characteristics and Requirements for Point-to-Point Equipment and Antennas; Part 4-2: Antennas; Harmonized EN Covering the Essential Requirements of Article 3.2 of the R&TTE Directive," version 1.5.1, 2010.

[8] ARIB, "Millimeter-Wave Data Transmission Equipment for Specified Low Power Radio Station (Ultra High Speed Wireless LAN System)," STD-T74, version 1.1, 2005.

[9] ACMA, "Radiocommunications (Low Interference Potential Devices) Class Licence 2000," amendment 2009.

[10] Ministry of Economic Development, "Radiocommunications Regulations (General User Radio Licence for Short Range Devices) Notice 2007," 2007.

[11] Moraitis, N., and P. Constantinou, "Indoor Channel Measurements and Characterization at 60 GHz for Wireless Local Area Network Applications," *IEEE Trans. Antennas and Propagation*, Vol. 52, No. 12, 2004, pp. 3180–3189.

[12] Gilb, J. P. K., and S. L. Li, "Millimeter Waves for Wireless Networks," in *Millimeter Wave Technology in Wireless PAN, LAN, and MAN*, S-. Q. Ziao, M-. T. Zhou, and Y. Zhang, (eds.), Boca Raton, FL: CRC Press, 2008.

[13] Wells, J. A., "MM-Waves in the Living Room: The Future of Wireless High Definition Multimedia Connectivity," *Microwave Journal*, Vol. 52, No. 8, 2009, pp. 72–84.

[14] WirelessHD Consortium, "WirelessHD Specification Overview," Version 1.0a, 2009.

[15] 802.15.3c-2009, "IEEE Standard for Information Technology—Telecommunications and Information Exchange Between Systems—Local and Metropolitan Area Networks—Specific Requirements, Part 15.3: Wireless Medium Access Control (MAC) and Physical Layer (PHY) Specifications for High Rate Wireless Personal Area Networks (WPANs) Amendment 2: Millimeter-Wave-Based Alternative Physical Layer Extension," 2009.

[16] ECMA-387, "High Rate 60 GHz PHY, MAC and HDMI PAL," 2008.

[17] ISO/IEC 13156, "Information Technology—Telecommunications and Information Exchange Between Systems—High Rate 60 GHz PHY, MAC and HDMI PAL," 2009.

[18] Niknejad, A. M., "Siliconization of 60 GHz," *IEEE Microwave Magazine*, Vol. 11, No. 1, 2010, pp. 78–85.

[19] Gilbert, J. M., et al., "A 4-Gbps Uncompressed Wireless HD A/V Transceiver Chipset," *IEEE Micro*, Vol. 28, No. 2, 2008, pp. 56–64.

[20] Pinel, S., et al., "60GHz Single-Chip CMOS Digital Radios and Phased Array Solutions for Gaming and Connectivity," *IEEE Journal on Selected Areas in Communications*, Vol. 27, No. 8, 2009, pp. 1347–1357.

[21] Marcu, C., et al., "A 90 nm CMOS Low-Power 60GHz Transceiver with Integrated Baseband Circuitry," *ISSCC Dig. Tech. Papers*, San Francisco, CA, February 8–12, 2009, pp. 314–315.

6

Multi-Gigabit 70/80-GHz and Higher Millimeter-Wave Radios

6.1 Introduction

The history of the 70 GHz and higher millimeter-wave bands goes back more than 30 years. The 71–76-GHz, 81–86-GHz, and 92–95-GHz bands were allocated for fixed wireless services by the ITU at WARC-79 (held in Geneva in 1979). The bands were first codified into the U.S. frequency allocation plan in January 1984. Minor modifications were made following WARC-92 (Malaga-Torremolinos, 1992), WRC-97 (Geneva, 1997), and WRC-2000 (Istanbul, 2000) to give the ITU allocations we have today.

Despite frequency allocations being available, no technical rules or regulations were in place to govern the bands, making them practically unusable. In September 2001, a petition was filed with the FCC requesting the establishment of service rules for licensed use of the 71–76-GHz and 81–86-GHz bands. At the time, the bands were reserved for government and military operations, as well as some satellite and radio astronomy applications. In September 2002, the FCC requested formal comments on this proposal. In October 2003, after consultation with industry and a number of wireless associations, the FCC ruled that 13 GHz of previously unused spectrum at 70, 80, and 90 GHz was available for high density fixed wireless services in the United States. This ruling was significant in a number of ways:

- The three allocations were (and still are) the highest frequencies commercially licensed by the FCC.

- The 13 GHz of spectrum opened up increased the amount of FCC approved frequency bands by 20% and represented 50 times the bandwidth of the entire cellular spectrum.

- A novel licensing scheme, allowing cheap and fast allocations to prospective users, was proposed.

- New markets, particularly for fiber extensions, were opened up, since the wide bandwidths permitted enabled gigabit-speed wireless communications with carrier-class performances over distances of more than a mile.

The speed with which the FCC acted highlights the enormous potential envisaged for these bands. Then-FCC Chairman Michael Powell heralded the ruling as opening a "new frontier" in commercial services and products for the American people [1].

This chapter focuses on the millimeter-wave bands of 70 GHz and higher and explores the band's characteristics and uses and how they enable gigabit per second and higher data rates. Since the main interest in this region is the 10 GHz of spectrum available in the 71–76-GHz and 81–86-GHz "E-band" allocation, the main focus of this chapter is on this part of the spectrum. Opportunities in the higher frequency bands at 90 GHz and above are covered later in the chapter, including proposals to use the upper-millimeter-wave bands for future 40-Gbps and 100-Gbps systems.

It should be noted that the 71–76-GHz and 81–86-GHz allocations are often referred to simply as the 70-GHz and 80-GHz bands, even though these limits are outside the allocations. This nomenclature makes the 70/80-GHz bands consistent with the 60-GHz naming convention. Somewhat less frequently, the bands are referred to as the 75- and 85-GHz bands. The 92–95-GHz frequency band discussed in detail later is widely referred to as the 90-GHz band, to be consistent with the 70- and 80-GHz bands, although it often (and more correctly) called the 94-GHz or the 95-GHz band.

6.2 Characteristics of the 70/80-GHz Bands

6.2.1 Frequency Bands and Channel Sizes

Although the millimeter-wave bands are allocated by the ITU, they still need to be incorporated into each country's national frequency allocation plans. Although the status of these bands varies by country, there is fortunately harmonization in the frequency band limits. Unlike the 60-GHz band, which has wide global variations, the 71–76-GHz and 81–86-GHz limits are consistently applied in those countries that permit operation in the bands. However, channelization does vary on a regional basis.

In the United States, the 70- and 80-GHz bands are allocated as two pairs of 5-GHz blocks extending from 71.0 to 76.0 GHz and 81.0 to 86.0 GHz. There are no subchannels defined, and users are permitted to use the full 5 GHz of bandwidth in each band, making 10 GHz of bandwidth available for FDD operation [2].

In Europe, the frequency allocation for the 70/80-GHz bands is managed by CEPT under ECC Recommendation 05(07) [3]. Unlike the unchannelized band plan in the United States, CEPT allocates nineteen 250-MHz channels within the band, plus a 125-MHz guard band at the top and bottom of each band. Several channel pairing methods are allowed. Figure 6.1 shows how individual 250-MHz channels can be paired together for an FDD operation with a 10-GHz transmit-receive (TR) spacing. Also shown is a sub-5 GHz TR spacing operation within the same 71–76-GHz or 81–86-GHz band. CEPT also permits TDD operation in the 70/80-GHz band.

Although the European band plan defines 250-MHz channels, it permits national regulators to allocate these channels with some flexibility. In particular, regulators are allowed to aggregate channels together to give larger assignments. Any number from 1 to 19 is permitted, effectively allowing channel sizes up to 4.75 GHz (the full 5-GHz band allocation less the two 125-MHz guard bands at the band edges). Individual countries are free to interpret and implement these recommendations at their own discretion, and so there are variations in how the channel sizes are set across Europe. In the United Kingdom, for example, full aggregation of the channels is permitted, making the full 4.75-GHz band available for individual users. Several other countries allow the same usage. In Switzerland, the full 71–76-GHz and 81–86-GHz spectrum is commercially available, but channel sizes are limited to 250 MHz, 500 MHz, 1.0 GHz, and 2.0 GHz only. Other administrations have only partially opened the bands, recognizing various legacy military and satellite bands which have been reserved in the past. The Czech Republic, for example, permits use of only the 74–76-GHz and 84–86-GHz portion of the band for commercial wireless communications. Several other European countries similarly prohibit commercial use of parts of the 70/80-GHz band allocation.

In addition to the United States and Europe, both Australia and New Zealand have also opened the 70/80-GHz bands for commercial applications. Australia has adopted a band plan following the United Kingdom (unchannelized use of the bands 71.125–75.875 GHz and 81.125–85.875 GHz) [4]. New Zealand uses the same frequency band limits, but only permits 250-MHz, 1.25-GHz, 1.75-GHz, and 2.25-GHz channels [5]. A number of other countries have also opened the 70/80-GHz bands for commercial use, while others allow instal-

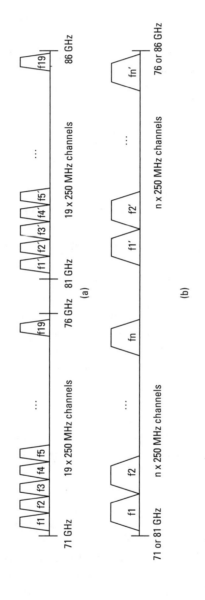

Figure 6.1 The 250-MHz channel allocations in the 70/80-GHz bands (a) for 10-GHz TR spacing and (b) within the same subband with less than 5-GHz TR spacing.

lations only for specific applications (for instance, for government or military use) with limited restrictions.

Currently the 70/80-GHz bands are not open or available for commercial use in major markets such as Canada, Mexico, South America, India, China, Japan, and most of Asia and Africa. Despite not being formally open, a number of 70/80-GHz links are installed in all these countries under special or temporary licenses. Regulators are also actively considering the bands. Unfortunately, it takes several years to move though the process to implement new frequency allocations (technical, economic and environmental studies, public consultations, considerations and responses, government voting, and rewriting and publishing of new telecom legislation). For this reason, it may be several years before full global adoption of the 70/80 GHz is in place.

6.2.2 Rules and Regulations

Rules and regulations for approving, operating, and managing 70/80-GHz millimeter-wave wireless equipment are very different in the United States and in Europe. Fortunately, most countries in the rest of the world follow the specifications from one of these two regions.

6.2.2.1 United States

In October 2003, the FCC added service rules for the 70/80-GHz bands to Part 101 regulations [6], making 70/80-GHz services consistent with all other fixed services in the United States. Since this time, the bands have been fine-tuned. Current FCC radio parameters, as required by Part 101 rules, are as shown in Table 6.1.

In general, these technical requirements are not strenuous and compliance is straightforward. Given the wide 5-GHz channels, system can operate

Table 6.1
FCC Part 101 Rules for 70/80-GHz Radio Equipment

Parameter	Requirement
Maximum EIRP*	55 dBW (300 kW)
Maximum transmitter power	35 dBm (3W)
Maximum transmitter power spectral density	150 mW per 100 MHz
Minimum antenna gain	43 dBi
Minimum spectral efficiency	0.125 bps/Hz
Maximum out of band emissions	−13 dBm

* For antenna gains of less than 50 dBi (but greater than or equal to 43 dBi), the maximum EIRP is reduced 2 dB for every 1 dB of antenna gain. Thus, the maximum EIRP with minimum antenna size of 43 dBi is 41 dBW.

mid-band (73.5 GHz and 83.5 GHz) and can use occupied bandwidths of up to 5 GHz in which to transmit gigabit per second data. At the band edges (71 GHz, 76 GHz, 81 GHz, and 86 GHz), relatively light attenuation is needed to meet the −13-dBm out of band emissions. Thus, compliance to the FCC radio parameters is relatively easy.

Despite the relatively gentle radio requirements, FCC antenna requirements are a lot more stringent. Tight radio pattern envelope (RPE) requirements are defined in Part 101 §101.115(b)(2), which are detailed in Figure 6.2. Very highly directional antennas are required, with low cross-polarization leakage and minimal sidelobes and back lobes permitted. These strict requirements are deliberate to ensure that the 70/80-GHz emissions radiate as "pencil beams" from these antennas. From a system perspective, these highly focused beams enable straightforward interference analysis and permit spatial (geographical) coordination of links, both of which are key elements for the novel 70/80-GHz licensing scheme implemented in the United States and discussed in Section 6.2.3.1.

Since the 70/80-GHz bands are managed under Part 101 rules similar to all other licensed microwave equipment, equipment authorization prior to marketing and sales is the same for microwave radios, as discussed in Section 4.2.2.

6.2.2.2   Europe

In Europe, the technical rules for 70/80 GHz are significantly more complex and stringent than those defined by the FCC. For radio device type approval, the same specification subpart that is applicable to 60 GHz is applicable to 70/80-GHz equipment. This is ETSI EN 302 217 Part 3 for equipment operating in frequency bands where uncoordinated (unlicensed) deployment might be applied

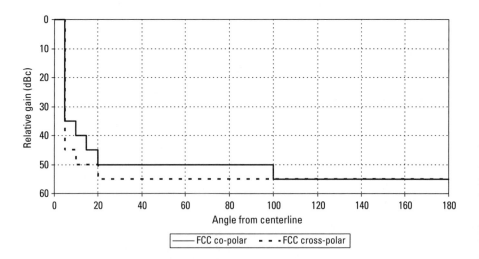

Figure 6.2   FCC 70/80-GHz RPE requirements.

[7]. The same EMC and safety specifications applicable to the lower-frequency microwave and 60-GHz millimeter-wave bands apply also to the 70/80-GHz bands. Similar to microwave and 60-GHz devices, all 70/80-GHz equipment needs to carry a CE mark, obtained by formal compliance against this RF type approval and the relevant EMC and safety specification before devices can be sold in Europe.

European 70/80-GHz Transmitter Requirements

EN 302 217-3 contains the test methodology and required technical parameters for 70/80-GHz equipment. The suite of requirements is similar to those for 60 GHz, although the specification limits are very different. For a 70/80-GHz transmitter, EN 302 217-3 details the following requirements, against which compliance is mandatory:

- Output power and EIRP;
- Output power tolerance;
- EIRP spectrum mask;
- Spurious emissions;
- Frequency tolerance;
- Minimum antenna gain.

These requirements have been in development for a number of years and were finalized in July 2009. The current released specifications are explained next.

ETSI defines a complex set of relationships between output power, EIRP and antenna gain. A maximum EIRP of 55 dBW (300 kW) is permitted, with a minimum antenna size of 38 dBi allowed [approximately an 8 inch (20 cm) parabolic antenna]. Both maximum EIRP and output power vary with antenna gain, as shown in Figure 6.3. Note that automatic transmit power control (ATPC) is allowed, although not mandated. For non-ATPC equipment, the maximum allowed output power is 30 dBm (1W) for systems with antenna gains of 45 to 55 dBi, and less for both smaller and larger antennas. Permitted maximum and minimum antenna gain also vary with output power, as shown in Figure 6.4. For equipment operating with the highest allowable output power (30 dBm), only antennas with between 45- and 55-dBi gain are permitted.

The ETSI 70/80-GHz transmitter power spectral density emissions mask is shown in Figure 6.5. Similar to the transmitter masks defined for microwave equipment, the 70/80-GHz mask is also measured at the transmitter output (i.e., at the antenna input port). However, unlike the microwave masks, it is an absolute mask rather than a relative mask. As such, lower output power 70/80-GHz

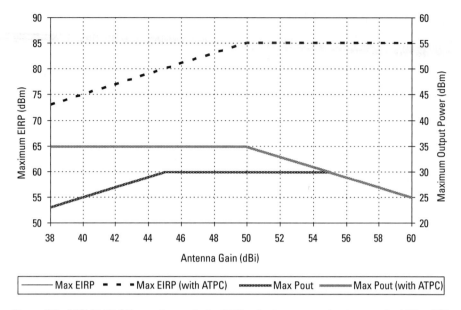

Figure 6.3 ETSI 70/80-GHz requirements for EIRP, output power, and antenna gain. (*After:* [7].)

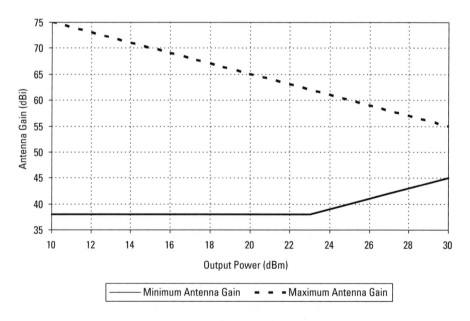

Figure 6.4 ETSI 70/80-GHz requirements for maximum and minimum antenna gain. (*After:* [7].)

devices have more margin to the emissions mask than higher power devices (contrary to microwave devices where the mask is measured relative to the output power of all transmitters).

There are several specific criteria to note with this mask in order to make valid measurements. First, the mask depends on the channel size used. This could be the single 250-MHz channel or a number of aggregated channels. This channel size determines the frequency offset values for the mask. Second, the power limits are defined as power densities, meaning that the limit depends on the bandwidth of the emission in addition to its power level. Third, the mask is not inclusive of frequency tolerance, which the tester needs to account for. Fourth, the mask applies to two classes of equipment, depending on the modulation format employed. A Class 1 system is one with a two-state modulation, such as FSK or BPSK. Class 2 and above are those with four-state or higher modulation (e.g., QPSK and above). The additional margin at band center for Class 1 systems is to permit possible carrier leakage associated with the less efficient Class 1 modulations. Finally, the maximum 250% limit applies only for aggregate channel sizes of 500 MHz or less. For larger channels, this limit is 150% + 500 MHz.

Another area where the ETSI transmitter limits require some explanation is out-of-band (OOB) emissions. Here three requirements need to be satisfied, each applicable to different areas of the frequency spectrum. The first OOB emissions requirement applies to transmitter mask measurements that fall outside the 71–76-GHz or 81–86-GHz frequency limits. Since the transmitter mask is defined over frequencies of 2.5 times the channel size, systems operating with wide aggregated channels or close to the band edges will likely have some of the mask fall outside these band edges. Under this circumstance, the OOB emissions outside the band edges up to 2.5 times the channel size are limited to a maximum of −55 dBW/MHz.

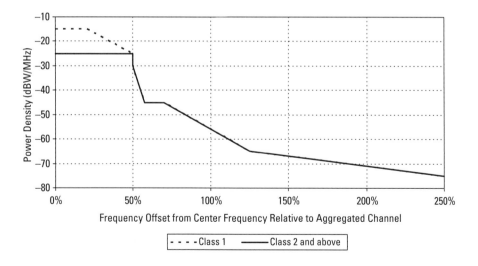

Figure 6.5   ETSI 70/80-GHz requirements for power spectral density. (*After:* [7].)

The second OOB emission criterion is specifically for the frequency band 86 to 92 GHz. These frequency bands are allocated in Europe for passive satellite services such as the Earth Exploration Satellite Service. To ensure protection for these services, ETSI requires the additional limits shown in Figure 6.6 for 86–92-GHz emissions. These limits cause a sharp discontinuity in emissions requirements at the 86-GHz band edge. For systems operating with wider bandwidths or in the higher ends of the bands, additional filtering is required to ensure compliance.

The final OOB requirement is for any spurious emissions that fall outside the two previous OOB criteria. This is defined, similar to any other point-to-point link, by CEPT ERC/Rec 74-01 [8]. For equipment operating above about 23 GHz, this requires that all spurious emissions outside the OOB limits defined above be less than −30 dBm. Compliance is required from 30 MHz to the second harmonic of the operating frequency.

Although it is not an essential requirement, ETSI provides guidance on the typical throughput expected for 70/80-GHz radios. This is quantified via a radio interface capacity (RIC), which is effectively a measure of spectral efficiency. ETSI provides the following RIC values, which are related to modulation, as defined in Table 6.2.

The concept of RIC is often found to be confusing. To give an example of how this is used, consider a QPSK 70/80-GHz radio operating with full gigabit Ethernet data throughput. This is a Class 2 radio (QPSK is a four-state modulation) with an over-the-air data rate of approaching 1.3 Gbps (1.25 Gbps for the 8B/10B coded gigabit Ethernet signal plus an additional overhead for FEC, NMS, and auxiliary data traffic). From Table 6.2, such a system can be permitted in any channel size of up to 1,000 MHz. The value of 1 GHz cannot be exceeded

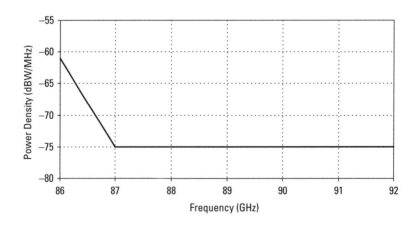

Figure 6.6   ETSI 70/80-GHz requirements for 86–92-GHz emissions. (*After:* [7].)

Table 6.2
ETSI 70/80-GHz Requirements for Radio Interface Capacity

Aggregate Channel (MHz)	Minimum RIC Values (Mbit/s)					
	128 -State Modulation	64 -State Modulation	32 -State Modulation	16 -State Modulation	4 -State Modulation	2 -State Modulation
250	1,000	900	750	600	300	150
500	2,000	1,800	1,500	1,200	600	300
750	3,000	2,700	2,250	1,800	900	450
1,000	4,000	3,600	3,000	2,400	1,200	600
1,250	5,000	4,500	3,750	3,000	1,500	750
1,500	6,000	5,400	4,500	3,600	1,800	900
1,750	7,000	6,300	5,250	4,200	2,100	1,050
2,000	8,000	7,200	6,000	4,800	2,400	1,200
2,250	9,000	8,100	6,750	5,400	2,700	1,350
2,500	10,000	9,000	7,500	6,000	3,000	1,500
2,750	11,000	9,900	8,250	6,600	3,300	1,650
3,000	12,000	10,800	9,000	7,200	3,600	1,800
3,250	13,000	11,700	9,750	7,800	3,900	1,950
3,500	14,000	12,600	10,500	8,400	4,200	2,100
3,750	15,000	13,500	11,250	9,000	4,500	2,250
4,000	16,000	14,400	12,000	9,600	4,800	2,400
4,250	17,000	15,300	12,750	10,200	5,100	2,550
4,500	18,000	16,200	13,500	10,800	5,400	2,700
4,750	19,000	17,100	14,250	11,400	5,700	2,850

Source: [7].

since the next largest channel size—1.25 GHz—requires a minimum capacity of 1,500 Mbps. Thus, an effective restriction on spectral efficiency has been made.

One practicality of RIC is that it clarifies the channel sizes used when measuring against the transmitter mask. The desire is to define channel sizes as large as possible, to simplify compliance against the steeper roll offs in the emission mask. However, the RIC places a limitation on how wide an occupied bandwidth can be defined.

European 70/80-GHz Receiver Requirements

The ETSI specification EN 302 217-3 also defines the receiver requirements for 70/80 GHz radios. These requirements include:

- Bit error rate (BER) as a function of receiver input signal level (RSL);
- Cochannel interference sensitivity;
- Adjacent channel interference sensitivity;
- CW spurious interference.

Unlike the transmitter requirements, these receiver limits are indicative. Therefore, it is not strictly necessary for equipment to comply with these limits, although properly designed devices should at least meet the given parameters. The 70/80-GHz receiver requirements are straightforward and consistent with both the microwave and 60-GHz millimeter-wave specifications. They are therefore not considered further here.

European 70/80-GHz Antenna Requirements

Similar to all other European microwave and 60-GHz devices, EN 302 217 Part 4-2 defines the antenna characteristics for 70/80-GHz systems [9]. For 70/80-GHz equipment, three classes of antennas are defined—class 2, 3, and 4—each of increasing antenna directivity. (A class 1 70/80 GHz antenna is also defined in Part 4-1 of this specification series, but its use is not permitted within the EU R&TTE framework.) Figure 6.7 shows the defined radiation pattern envelope (RPE) requirements for 70/80-GHz radios for the three classes of antenna. The ETSI specification also provides guidance that 27 dB of antenna cross-polarization distortion (XPD) should be targeted and reiterates that at least 38 dBi of antenna gain is required.

6.2.2.3 Australia

The Australian Communications and Media Authority (ACMA) manages the radio spectrum in Australia. In 2007, the ACMA authorized the used of the 70/80-GHz bands, with an open frequency plan similar to the United Kingdom and technical rules matching the then-current European draft rules [4]. Although the European rules have since been modified significantly, the ACMA specifications have not been updated to reflect current European usage.

Specifically, ACMA rules for the 70/80-GHz bands require a minimum antenna gain of 43 dBi, with a maximum permitted EIRP of 45 dBW. The maximum transmitter power is 30 dBm (1W). An RPE equivalent to the FCC antenna requirements is also specified.

6.2.2.4 New Zealand

In New Zealand, the radio spectrum is managed by the Radio Spectrum Management Department of the Ministry of Economic Development. In 2009, the 70/80-GHz bands were opened, with a frequency plan that follows the European

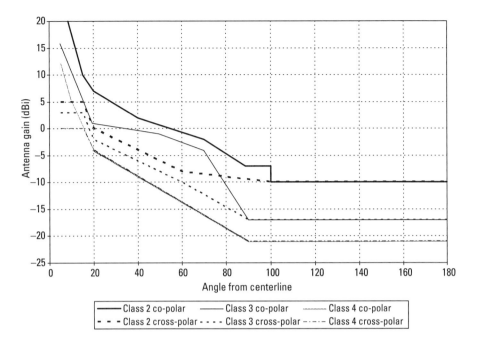

Figure 6.7 ETSI 70/80-GHz RPE requirements. (*After:* [9].)

band limits and technical rules that follow United States limits [10]. Specifically, a minimum antenna gain of 43 dBi is required, with a maximum permitted EIRP of 55 dBW. The maximum transmitter power is 3W with a maximum spectral density of 150 mW/100 MHz. The antenna RPE requirement also follows the FCC requirements.

6.2.3 Comparison of FCC and ETSI Rules

The United States' FCC rules and the European ETSI rules form the basis for technical regulations around the world. As shown in the pervious section, the two differ significantly, with the ETSI requirements generally being much more stringent. Comparisons can be made using the information previously presented. Two areas where FCC and ETSI specifications are very different and comparisons are not intuitive are the transmitter emissions mask and antenna characteristics. This section explores these further.

6.2.3.1 Power Spectral Density

Figure 6.8 shows a power spectral density measurement for a 20-dBm output power 70/80-GHz radio transmitting 1 Gbps of Ethernet data. The radio employs QPSK (class 2) modulation to keep the transmitted signal within a 1-GHz bandwidth (four aggregated ETSI 250-MHz channels). The transmitter is at the

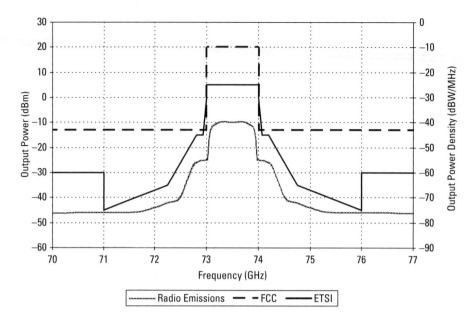

Figure 6.8 Example 70/80-GHz transmitter emission showing compliance to both FCC and ETSI power spectral density mask limits. (Note: Measurement bandwidth of 1 MHz.)

center of the 71–76-GHz band. Also shown are the FCC and ETSI power spectral density masks to illustrate the difference in the two requirements.

The FCC requirement is that emissions can be no higher than −13 dBm outside the band edges. Figure 6.8 shows that this compliance is straightforward, since the 20 dBm/1 GHz transmission, when viewed on a spectrum analyzer with an appropriate measurement bandwidth will be spread out to about that level. Compliance against the ETSI requirement is more difficult to establish, as the measured emission has to be converted into units of dBW/MHz to compare against the mask. Since the transmission has 20 dBm (−10 dBW) output power and occupies 1 GHz (30 dB MHz) bandwidth, it has a power spectral density of −40 dBW/MHz at the center frequency. Thus, the emission has a 15-dB margin against the −25-dBW/MHz mask specification at the band center, and the emission can be plotted against the mask. The ETSI OOB emissions should also be considered. In this case, emissions have to be no more than −30 dBm outside the band edges.

The FCC and ETSI antenna requirements are also very different. ETSI permits a minimum antenna size of 38 dBi [approximately an 8-inch (20-cm) parabolic antenna], which is smaller than the FCC minimum antenna requirement of 43 dBi [approximately a 12-inch (30-cm) parabolic antenna]. The RPE requirements are very different. Although ETSI defines RPE against absolute antenna gain and FCC defines RPE versus relative antenna gain, the two requirements can be scaled and directly compared. Figure 6.9 shows such a comparison

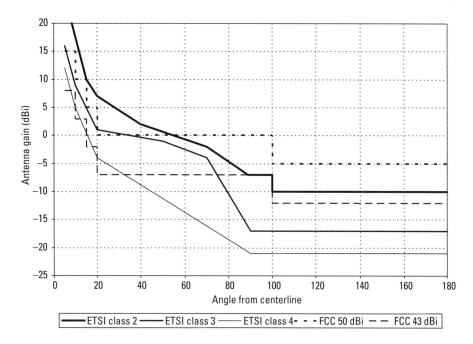

Figure 6.9 Comparison of ETSI and FCC 70/80-GHz RPE requirements.

for the three classes of ETSI antenna and for both a 50-dBi and 43-dBi FCC antenna. For forward angles up to about 40° from the antenna boresight, the 50-dBi FCC antenna specification lies about midway between the class 2 and class 3 patterns, while the 43-dBi FCC requirement follows the class 4 ETSI specification closely. However, the ETSI specifications are significantly more restrictive when considering the reverse emissions (90°–180°), with much more stringent front-to-back ratios required. Currently, a 43-dBi [approximately 1-foot (30-cm) parabolic] antenna that meets FCC requirements will meet ETSI class 2 requirements, but not necessarily class 3 or 4. A 50-dBi [approximately 2-foot (60-cm) parabolic] antenna that meets FCC requirements is not guaranteed to meet any of the ETSI requirements.

6.2.4 Licensing

Licensing for the 70/80-GHz band differs from that required for the microwave and 60-GHz bands. In the microwave bands, wireless systems are widely deployed, and available spectrum is in great demand. Since spectrum is a finite resource and generally demand exceeds supply, licensing the spectrum is an economically efficient way of allocating a scarce resource to those most able to utilize the medium. In the 60-GHz band, however, where very wide bandwidths are available, and high atmospheric absorption limits link distances making frequency reuse easier, spectrum scarcity is not such an issue. Sixty GHz is therefore generally given license exempt status, and operators are free to install devices and links without the need to register units or obtain licenses. The characteristics of the 70/80-GHz band fall between these two cases, so a different licensing approach is appropriate. Such an approach differs in each country that permits 70/80-GHz systems.

6.2.3.1 United States

In 2005, the FCC implemented a novel "light licensing" scheme for the 70/80-GHz bands that enables links to be cheaply and quickly coordinated and registered over the Internet [2, 11]. In deciding to go this route, the FCC considered two unique characteristics of the 70/80-GHz bands, reinforced through the way the bands have been implemented and managed, that are not experienced by conventional lower frequency devices. First, since the 71–76-GHz and 81–86-GHz frequency bands are configured as two single channels with no subbands or channelizations, traditional frequency planning does not need to be considered. Second, the high operational frequencies of 70/80-GHz systems, coupled with the high directivity antenna requirements, means that systems communicate via highly focused "pencil beam" transmissions. Operators are thus able to configure links close to one another without interference concerns. With no frequency coordination and much simplified interference analysis, traditional link licensing schemes are not necessary. For these reasons, the "light license" approach described next was proposed and adopted.

In order to file for a 70/80-GHz license, the applicant needs to first register to be a nationwide licensee with the FCC. This application takes just a few weeks and carries a small filing fee (currently $645). Once approved, the applicant is issued a call sign and can register any number of individual 70/80-GHz links in the United States and its territories. As of February 2010, there were 62 such 70/80-GHz license holders in the United States.

The application for an individual 70/80-GHz link license is via one of three FCC-appointed database managers: Comsearch, Micronet Communications, or Frequency Finder. These companies manage the link registration process and maintain a shared, central 70/80-GHz link database. Link licensing involves simply logging onto any one of the three database manager's Web sites, entering the user's call sign, and inputting a few technical link parameters (for example, longitude, latitude, and height of each end site, antenna size, transmission power, and some additional standard wireless equipment parameters).

To register and license the link, the database manager then undertakes a four-step analysis. First, a time and date stamp is assigned to the application, to resolve any future time-based conflicts. Second, an interference analysis is conducted, whereby the link characteristics are automatically mapped against other closely located links to identify any possible interference issues. Third, a check is made to ensure the link does not violate any of three specific FCC-imposed rules. These include risk of international cross-border transmission (special coordination is required for links within 35 miles of either the Canadian or Mexican borders, whose antenna points towards that border within a 200° sector, or within 5 miles if the antenna points away from the border within a 160° sector), proximity to radio astronomy quiet zones (nine radio astronomy sites have 150-km radius protection zones, and 10 further sites have 25-km protection zones within which special coordination is required), or violation of special antenna rules. As a final step, link parameters are passed to the National Telecommunications and Information Administration (NTIA) for final interference analysis against 28 defined military stations and a further number of undisclosed military and government operations.

After successful analysis of the proposed link, the applicant is advised that link registration is complete and electronic payment for the license is requested. Given there are three competing database managers, market forces keep processing fees low. Currently, individual link registrations fees are just $75. If there is an issue flagged with the license analysis, the applicant is requested to manually file a registration with the FCC. This further process could take several weeks, but does not incur any additional FCC fees.

The issued 70/80-GHz link license is valid for 10 years, and licensees are required to build and commission the link within the first year, or else the license is revoked. As of February 2010, there were almost 2,000 registered 70/80-GHz links in the United States.

6.2.3.2 United Kingdom

In 2007, the United Kingdom opened the 70/80-GHz bands with a light licensing process similar to that of the United States [12]. First, the applicant has to submit form OfW 383 to Ofcom, the national regulator, to register as a nationwide licensee. After approval, the licensee can apply for any number of individual link licenses. These are done via submitting form OfW 368 to Ofcom, containing details of equipment parameters and site information. A fee of £50 (approximately $75) is required, which also covers the first year's license. After this, an ongoing £50 per year fee is required. Ofcom maintains and publishes a database of registered 70/80-GHz links, which is updated weekly. As of February 2010, 32 links in the 70/80-GHz bands had been licensed in the United Kingdom.

The U.K. and U.S. systems differ in two ways: the U.K. system is currently a manual process, and the U.K. licensing process does not perform an interference analysis or provide interference protection. Ofcom requires links to be self-coordinated by the applicants and deployed sensibly and with consideration to other nearby users. Link licenses, however, are time- and date-stamped, and, in the unlikely event that interference was to occur, Ofcom requires priority to be given to the first link installed. The more recent link is required to be reconfigured or modified to eliminate the interference.

6.2.3.3   Rest of Europe

The 70/80-GHz bands are managed differently in various parts of Europe. Many countries have determined that 70/80-GHz services should be license-exempt. In the Czech Republic, for example, links can be freely deployed without the need to register or coordinate the equipment. Other countries have determined that the 70/80-GHz links should be treated the same as conventional microwave links. Ireland, for example, permits 70/80-GHz operation under the same license conditions as all other point to point links at 1 GHz and above. License fees in Ireland are currently set at €952.30 (approximately $1,300) per year. Other countries in Europe are seen charging as much as €3,000 (approximately $4,100) per year for 70/80-GHz licenses.

6.2.3.4   Middle East

Several nations in the Middle East permit 70/80-GHz operation. The United Arab Emirates (UAE) permits high capacity millimeter-wave links under its traditional point-to-point link licensing rules. License fees are set by the UAE Telecommunications Regulatory Authority (TRA), which applies a fixed formula that takes into account the frequency, power, and bandwidth of the link, placing weighted multipliers to encourage spectral efficiency and frequency reuse. A typical 70/80-GHz radio will incur an annual license fee of about 4,500 dirhams (approximately $1,200). In Bahrain, 70/80-GHz links are also permitted under regular radiocommunications policy. Here the TRA of Bahrain applies a simple license fee determination of 1% of gross turnover arising from the activities associated with that license.

6.2.3.5   Australia and New Zealand

The ACMA permits 71–86-GHz links to be operated in Australia under a "light licensed" process that is essentially identical to that used in the United Kingdom, with a license fee set at AU$187 (approximately $170) per year. As of February 2010, there were almost 150 commercial 70/80-GHz links registered in Australia. In New Zealand, the Ministry of Economic Development allows 70/80-GHz systems under the regular licensing process. Links have to be registered in the centralized Register of Radio Frequencies and certified by an approved radio engineers before commissioning. License fees are currently set at NZ$200 (approximately $140) annually. This is the same fee schedule as for all other fixed radio links.

6.2.5   System Characteristics

The 70/80-GHz bands are available for use in many of the major global markets, and the very wide channel bandwidths allowed permit very high data rates to be transmitted. Even with the simplest of modulation schemes, these bandwidths are sufficient to easily transmit a gigabit per second of data. With a more spectrally efficient modulation, full duplex data rates of 10 Gbps can be reached. With direct data conversion and low cost diplexers, relatively simple and thus cost-efficient and high reliability radio architectures can be realized.

One key limitation of 70/80-GHz transmission is link distance. As shown in Chapter 7, the distance over which millimeter-wave radios can transmit is strongly determined by the rain characteristics in that region. In dry climates, commercial 70/80-GHz links have been installed with distances over 10 km. However, over such distances, links would be expected to experience outages during even light rain showers. In most parts of the world, links are practically limited to about 1.5 to 2 km (1 to 1.5 miles) with 99.999% weather availability (i.e., the radio will operate statistically for 99.999% of the time, or only be down for 0.001% of the time, or approximately 5 minutes per year), and 5 to 7 km (3 to 4.5 miles) with 99.9% weather availability (down for about 8 hours per year).

6.3   70/80-GHz Market Applications

6.3.1   Cellular and WiMAX Backhaul

As shown earlier in Section 4.3.1, PTP wireless has played a significant role in connecting together cell sites and cellular base stations for many years. Approximately 50% of the world's 4 million cell sites are currently connected using microwave wireless. Due to significant network loading, caused by increased data rate usage by mobile broadband phones and other devices, carriers are deploying more 3G-advanced and 4G technologies such as LTE and WiMAX. These IP-based networks, in turn, require increasingly higher capacity backhaul.

It has already been shown how high capacity microwave radios can support such 3G and 4G standards (see Figure 4.1). Millimeter-wave radios at 70/80 GHz, which also offer high capacity transmission over distances of several kilometers, will play a similar role. Since 4G networks are still in their infancy, the question becomes how and where 70/80-GHz millimeter-wave radios will be installed in such networks.

There are a variety of ways in which next generation networks will be configured and similarly a number of ways in which high capacity millimeter-wave radios can be utilized within these networks. Figure 6.10 shows four such ways. Since 70/80-GHz radios have the capability of transmitting data rates in excess of 1 Gbps, their strength is in aggregating traffic from various cell sites. In a traditional hub-and-spoke arrangement, as shown in Figure 6.10(a), high capacity

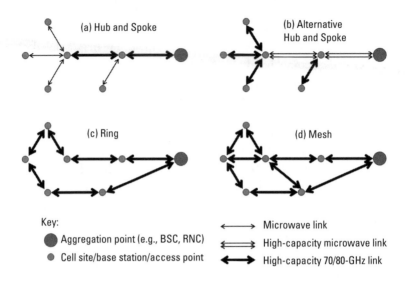

Figure 6.10 (a–d) Four possible network topologies for next generation networks.

microwave radios supporting data rates of several hundred megabits per second are used to connect the outlying base stations, and 70/80-GHz radios are used to interconnect these and carry the aggregate traffic at gigabit per second speeds back to the BSC/RNC network controller. An alternative view is shown in Figure 6.10(b). Since there are many more outlying cell sites around the edges of the network, spectrum allocation and frequency planning becomes more problematic in these areas. Since 70/80-GHz operates in unused spectrum and offers light licensing in most countries, whereby interference protection can be guaranteed much quicker and cheaper than for microwave radios, licensed millimeter-wave devices will be more cost-effective at the edge of the network. The same argument is true in dense metropolitan areas. In this scenario, millimeter-wave radios will provide the bulk of the edge connectivity, and high capacity microwave radios will provide the aggregated backhaul, since microwave offers the advantages of longer-distance transmissions. A third network topology that is becoming widely used is a ring configuration, as shown in Figure 6.10(c). Ring topologies are common in fiber optic networks and are becoming more popular in wireless networks. In a ring, each node has a dual path to every other node, making the whole network more robust than a hub and spoke configuration. However, if one link becomes congested, the entire ring may be affected. For this reason, rings need to utilize high capacity links to ensure that congestion is minimized, adding to the appeal of 70/80-GHz devices for this application. Finally, a fourth scenario is that next generation networks will be installed as mesh networks rather than hub-and-spoke or ring configurations, as in Figure 6.10(d). In a mesh, each link has to carry an increased traffic load from multiple base stations. In this case, the higher data carrying capacity of 70/80-GHz radios will lend itself better to mesh networks, since millimeter-wave radios offer higher throughput and more economic licensing. All four network scenarios are feasible, with many more pros and cons to each. However, the potential role in which high capacity millimeter-wave links can play in each topology can be seen.

It is worth noting that base station separation distances will play a role in determining the interconnection technology used. Cell sizes have been steadily dropping over the years. Figures from the United States show that in the 1980s, cell sites were routinely deployed with cell radiuses averaging a 20-mile (35-km) radius. In the 1990s, this dropped to about 3 miles (5 km). In the 2000s, distances between United States cell sites dropped to about 1 mile (1.6 km) in urban areas and about twice this in suburban areas. Next generation networks will see installations with much closer base station installations. Thus, the limited transmission distance of 70/80 GHz will not be a major impediment to its adoption as a choice for cellular or WiMAX backhaul.

6.3.2 Remote Radio Heads/Distributed Antenna Systems

A cellular application where 70/80-GHz wireless will play a significant role is supporting Distributed Antenna Systems (DAS). DAS is an application whereby cellular coverage is improved by replacing a single antenna connected to a base station with a number of smaller spatially distributed antennas, each connected to the same single base station. By splitting the transmitted power among several separated antenna elements, coverage over a given area can be provided with reduced power and improved reliability. DAS operates on the concept that distributed antennas waste less power in overcoming penetration and shadowing losses. The more frequent line of sight leads to reduced fade depths and less delay spreads. In addition, DAS allows antennas to be placed in areas hard to reach with traditional systems, such as in areas of shadowing or inside buildings.

Often DAS is employed with remote radio heads (RRH), whereby an active RF transceiver (a "radio head") is located remotely with the distributed antenna to improve the performance of the system. All the RRH are connected back to a base station via a common medium, usually fiber optic cable. Either CPRI or OBSAI optical interfaces are used to provide the interconnection protocol between the base station modem and the remote radio/antenna equipment. Both CPRI and OBSAI are described in detail in Section 3.3.4.4.

Figure 6.11 shows a typical RRH/DAS application. Since the CPRI and OBSAI data rate can be at gigabit per second speeds ($m \times 768$ Mbps, where m = 1, 2, or 4 for OBSAI, and $n \times 614.4$ Mbps, where n = 1, 2, 4, 5, 8, and 10 for CPRI), there is an opportunity for high capacity 70/80-GHz wireless to replace the usual fiber connection and provide the interconnection between the base station modem and the remote radio/antenna equipment.

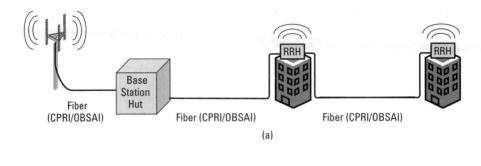

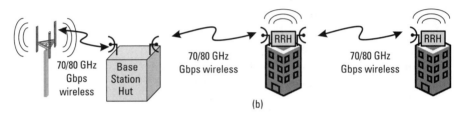

Figure 6.11 (a) Traditional RRH/DAS application using CPRI or OBSAI fiber and (b) alternative configuration using multi-gigabit per second 70/80-GHz wireless.

6.3.3   Enterprise Connections

Similar to microwave and 60-GHz radios, 70/80-GHz radios can be used to provide high data rate point-to-point enterprise connections between businesses, schools, hospitals, campus buildings, and any other enterprise. This application was discussed in Section 4.3.2.

6.3.4   Fiber Extensions/Backup

For many years, fiber optic networks have been aggressively rolled out around the world to provide national and international backbone infrastructure networks. Despite there being a large amount of installed fiber, many buildings, businesses, and other enterprises do not have access to such fiber connections. Studies in 2010 have shown that although the majority of large enterprise sites are fiber-connected, small to medium business sites are vastly underserved. In the United States, only 23% of buildings with 20 or more employees have access to fiber. In Europe the number is smaller at 15%. For those buildings with no fiber connections, businesses have to rely on leasing wireline circuits from the incumbent carrier, competitive providers, or local cable operator for services. Such services are much slower than fiber, and can run as high as thousands of dollars per month.

There is therefore an unserved need for short-haul wireless connectivity in the last mile. Gigabit 70/80-GHz systems provide fiber-like speeds at distances of a mile or so, making them suitable for this application. Since pricing for fiber trenching can rise as high as $250,000 per mile in metropolitan areas and be prohibited in many locations, the use of 70/80-GHz links for extending fiber makes economic sense.

Similarly, because of its comparable data rate to fiber, 70/80-GHz wireless can be used to back up short-haul fiber links. In the United States, the 2005 Appropriations Act stipulated that telecommunications funding would not be provided for federally owned buildings that do not have a redundant telecommunications network with physically separate entry points from their primary network. Given that most telecommunications networks are via fiber optic cable or copper wiring to the basement or ground floor of the building, this law promotes the case for wireless telecom connectivity via different entry points, usually a roof or window. Wireless further strengthens network availability by having completely different outage mechanisms to wired connections. Wireline communications are subject to flooding, earthquakes, and construction breakages, whereas wireless is susceptible to weather conditions such as wind and rain. Given its ability to carrier fiber-like data rates, 70/80-GHz wireless is a leading contender for such fiber backups.

6.3.5 Other Applications

There are a number of other applications where 70/80 GHz will play a role. Some of these include:

- *Machine-to-machine connectivity for storage area networks:* Storage area networks (SAN) run at 1 Gbps or higher Fiber Channel rates. SAN interconnects and the remote backup of SAN at physically separate locations are possible using multi-gigabit 70/80-GHz radio links.

- *Portable and temporary links for high definition video or HDTV transport:* The high data rate carrying capacity of 70/80-GHz radios permits realtime uncompressed HDTV transmission for applications such as video surveillance, real-time high definition monitoring of sensitive sites, airports, or border control.

- *Military applications:* High data rate wireless has many defense applications including high fidelity image and sensor monitoring, HD video surveillance and perimeter control, secure high data rate communications, and large file transfers in hostile environments.

6.4   High Data Rate 70/80-GHz Radio Equipment

6.4.1   Commercial Equipment

There are a number of commercially available high capacity 70/80-GHz wireless devices in the market today. A typical system that provides 1-Gbps data capacity is shown in Figure 6.12. This radio consists of an all-outdoor configuration where all the interfaces, electronics, and millimeter-wave circuitry are enclosed in a single weather-sealed radio chassis. The radio operates FDD with a 10-GHz TR spacing and uses BPSK modulation to transmit the 1-Gbps data capacity in the 70/80-GHz bands. The radio is attached to a 2-ft (60-cm) directional antenna that provides 50 dBi of gain with 0.4° beamwidth, which enables the device to deliver a full 1 Gbps of data at distances of several kilometers with high weather availability.

6.4.2   System Architectures and Performance Tradeoffs

The internal architecture of a typical 70/80-GHz radio is shown in Figure 6.13. The operation of this is very similar to the concepts described earlier in Section 3.4. However, there are a number of elements in this configuration specific for 70/80-GHz radios.

Figure 6.12 Photographs of a typical gigabit per second 70/80-GHz radio connected to a 2-ft (60-cm) antenna. (*Source:* GigaBeam Corporation, 2010. Reprinted with permission.)

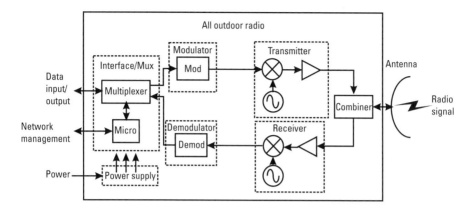

Figure 6.13 Block diagram of a typical gigabit per second 70/80-GHz radio architecture.

6.4.2.1 Interfaces

The gigabit Ethernet or other high data rate format traffic interface to a 70/80-GHz radio is usually by fiber via an optical short form pluggable (SFP) connector or a twisted-pair CAT 5 or 6 cable via an electrical RJ-45 connector. Similar to most other industrial wireless equipment, the power interface is usually at the telecommunications norm of −48-V DC. AC inputs are generally avoided as they add significantly to the approvals process due to additional EMC and safety compliance, and they also complicate international supply logistics because of varying global power standards.

6.4.2.2 Modems, Modulation, Bandwidth, and Throughput

Most commercial 70/80-GHz systems currently employ low-order, two-state modulation, typically OOK, FSK, or BPSK. Although these modulations are relatively easy to implement, they are not spectrally efficient. A typical gigabit Ethernet radio transmits data over the air at about 1.3 to 1.4 Gbps (the 1.0 Gbps data, plus 25% overhead for 8B/10B coding, plus an additional overhead for any FEC, network managing, and auxiliary traffic). Since a two-stage modulation has a theoretical modulation efficiency of 1 b/s/Hz, (3.2) and (3.3) showed that a typical system will have an occupied bandwidth of between 1.5 and 2.0 GHz once the modem Nyquist filter (α) and practical implementation inefficiencies are accounted for. Although this is not a concern for deployments in the United States, where the FCC permits a full 5-GHz channel for radio links, it is a concern for European adoption, whereby the notion of spectral efficiency is encouraged by regulators and is implicit (and sometimes explicit) in national rules.

For this reason, there is currently interest in deploying 70/80-GHz systems with higher-order modulations. Already some units with 4FSK and QPSK modulation are commercially available. These higher modulation systems offer better spectral efficiency, and can transmit full gigabit Ethernet traffic over the air in occupied bandwidths of 1 GHz or less. However, there are system penalties with employing these higher modulation systems. First, system complexity increases, which adds to design, production, and system costs. Second, receiver sensitivity reduces because of the higher signal-to-noise requirements of two-level modulation over single level systems. Finally, output power may be reduced because of the need to preserve linearity in the power amplifiers. Fortunately, 4FSK and QPSK are constant envelope modulations, so this is not a consideration, but output power reduction is a consideration for higher QAM modulations. The reductions in transmit power and receiver sensitivity both combine to shorten effective link distances.

There is therefore a tradeoff between modulation (and hence occupied bandwidth) and system performance and cost. ETSI rules suggest that modulations up to 128 QAM are envisaged in future 70/80-GHz systems (Table 6.2). As modulation increases, spectral efficiency improves, but system gain falls because of receiver sensitivity reduction and transmitter power backoff required to support these higher modulations. Some typical modulation and channel size tradeoffs for practical 1-Gbps systems are shown in Table 6.3. Note that the subject of link distances is covered in detail in Chapter 7.

For systems targeting much higher data rates, higher frequency modulations are a necessity. The highest published data rate for a 70/80-GHz link is 6-Gbps data rate, where researchers have built a multichannel demonstration link that uses 8 PSK modulation [13, 14]. To get to a true 10GbE transmission, an even higher modulation scheme is required. To transmit 10 Gbps of usable Ethernet data, an over-the-air data rate of about 13 Gbps will be required to account for protocol coding, radio coding, and link overhead. With realistic implementation losses, either a 32-QAM or a 64-QAM modulation (with 5 b/s/Hz and 6 b/s/Hz theoretical efficiencies, respectively) will be needed to compress the transmission into the permitted 5-GHz channels.

Another tradeoff in 70/80-GHz systems is the modem technology employed. In high capacity radios operating in the 6–40-GHz microwave bands, the modem functionality is implemented in a structured or programmable signal

Table 6.3
System Tradeoffs Required to Support Practical
1-Gbps Gigabit Ethernet Traffic

Modulation	Channel Size	Distance*
QPSK	750–1,000 MHz*	1 to 2 miles
16 QAM	500 MHz	0.5 to 1 mile
64 or 128 QAM*	250 MHz	Less than 1 mile

* Depending on implementation.

processing device such as an ASIC or FPGA. Such devices are able to cope with the required 56 MHz and smaller signal bandwidths. However, in the 70/80-GHz bands where gigabit-per-second capacities require bandwidths of 1 GHz or higher, digital processors and mixed signal devices (analog-to-digital and digital-to-analog converters) with sufficient performance are not available at commercially acceptable costs. For these reasons, large portions of 70/80-GHz modems are usually implemented with analog circuitry. For example, many wideband 70/80-GHz systems use a Costas carrier recover loop in the demodulator, which uses sensitive phase locked loop techniques to extract the required carrier. Component tolerances and manufacturing fluctuations place limits on achievable phase and amplitude balance, effectively placing a limit on how high this technique can be used. By focusing on the higher modulations and narrower channels, coherent all-digital techniques can be used, which can be implemented in low-cost, commercially available digital signal processing circuitry. Commercial FPGA-based modem designs are available for millimeter-wave transmission using the 250-MHz bandwidth, although at data rates much below 1 Gbps. Being all-digital, a straightforward roadmap to the higher QAM modulations and gigabit per second data rates is possible.

6.4.2.3 Millimeter-Wave Devices

There are very few commercial choices for 71–86-GHz millimeter-wave microwave monolithic integrated circuit (MMIC) devices and components. It was shown in Chapter 5 how CMOS silicon and Bi-CMOS SiGe devices are being used effectively at 60 GHz. However, such devices have not yet been extended commercially to higher frequency bands. The 71–86-GHz band is still the realm of traditional GaAs and InP devices. A number of systems use devices fabricated from the Northrop Grumman foundry, which uses a GaAs 0.15-μm pseudomorphic high electron mobility transistor (PHEMT) process to supply commercial and proprietary 70/80-GHz devices. This process has produced power amplifiers with output powers beyond 20 dBm at a 1-dB compression and low noise amplifiers with noise figures of 5 dB. Figure 6.14 shows die photographs of 71–86-GHz transmitter and reciver devices. A schematic for a full 71–86-GHz transceiver using these commercial off the shelf devices is given in [15].

6.4.2.4 Transmitter/Receiver Architecture

Another transceiver tradeoff is whether to utilize the two 5-GHz channels and transmit and receive in the 71–76-GHz and 81–86-GHz bands, respectively, with 10-GHz TR spacing, or whether to limit both transmission and reception to the lower 71–76-GHz band. Both options are permitted in the U.S. and European band plans. Several commercial systems have chosen to employ the single 71–76-GHz band approach. In this configuration, a TR spacing of much less than 5 GHz is used, with transmit and receive frequencies chosen close to the

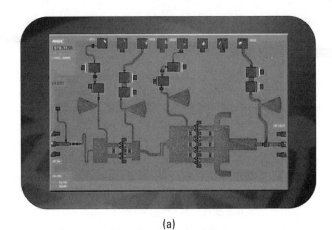

(a)

(b)

Figure 6.14 Die photographs of (a) a 2-stage 71–76 GHz power amplifier and (b) a 3-stage 71–86-GHz low-noise amplifer. (*Source:* Hittite Microwave Corporation, 2010. Reprinted with permission.)

top and bottom of the band to give as much frequency separation as possible. Transmission is usually on different polarizations and an orthomode transducer (often known as a polarization duplexer) is used to separate the transmit and receive channels in place of a regular diplexer combiner.

There are some advantages to operating in just the 71–76-GHz band. Transceiver device selection is easier, since MMICs have better performance at the lower frequency bands. Also, devices developed for other nearby millimeter-wave applications can be used (for example, the nearby 77-GHz band is being developed for automotive radar, a potential future mass market). Furthermore, by avoiding transmission in the 81–86-GHz band, the strict 86–92-GHz out-of-band emissions required by ETSI (Figure 6.7) are much less of a concern.

The disadvantage of limiting operation to just the 71–76-GHz band is that the maximum data rate supported is limited. Since a separation between transmit and receive is needed to avoid interference, much less than half the band is available for transmission. Systems are available that support full 1-Gbps transmission, but practically higher FDD data rates cannot be supported by equipment operating in the single subband.

6.4.2.5 Antennas

The final unique consideration for 70/80-GHz system tradeoffs is in the antenna. Characteristics are already well defined by both ETSI and the FCC, with stringent RPE and front-to-back ratios required. Minimum antenna gains are also required: 38 dBi in Europe and 43 dBi in the United States, limiting antenna diameters to no smaller than about 8 inches (20 cm) and 1 foot (30 cm), respectively. There is a tendency to use larger antennas to extend transmission distances or give more link margin (discussed more in Chapter 7). However, there are practical limitations resulting from the difficulty in aligning the narrow pencil beams for high gain antennas. Mounting poles and towers sway in the wind, and even buildings can twist as the sun heats up and moves across the structure. A 2-foot (60-cm) diameter 70/80-GHz antenna typically has a beamwidth of about 0.4° and 50-dBi gain. Such antennas are regularly employed for high capacity 70/80-GHz systems. However, commissioning any larger antennas requires skill and patience, and maintaining alignment is a challenge.

6.5 90-GHz and Higher Bands

When the FCC first announced service rules for the 70- and 80-GHz bands in 2003, they also included rules opening up and allowing operation in the 92–95-GHz frequency band. This allocation is widely referred to as the 90-GHz band and was also allocated to enable high data rate fixed wireless services.

Ninety-four GHz has been used for many decades as a military frequency. It is popular for covert applications because traditionally this part of the spectrum was not widely used, spectrum was available, and transmission characteristics are generally good (94 GHz is the minimum in the atmospheric attenuation profile; see Figure 2.5).

Like the 60-, 70-, and 80-GHz bands, the 90-GHz band offers a wide swath of spectrum to enable high data rate communications, as shown in Figure 6.15. However, the 90-GHz band is smaller than the other allocations and far more difficult to work with. The 90-GHz allocation is actually in two parts: 92.0–94.0 GHz and 94.1–95.0 GHz. Segmented into unequal portions of 2 GHz and 900 MHz and separated by a narrow 100-MHz exclusion band at 94.0–94.1 GHz to preserve existing military services, the asymmetric frequency allocations

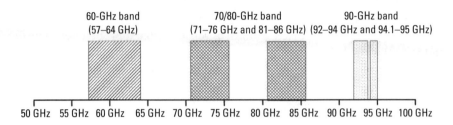

Figure 6.15 U.S. allocations for the 60-GHz, 70/80-GHz, and 90-GHz bands.

forces lower data throughputs and more complicated filtering schemes, both a deterrent to low-cost commercial deployments.

To further discourage use, the 90-GHz band is not commercially available in any other part of the world apart from the United States. In the United States, the 90-GHz band is light licensed and managed under the same Part 101 rules as the 70- and 80-GHz bands. At the time of this writing, there are no commercial high data rate systems available in the 90-GHz band.

In the United States, the FCC allocates spectrum up to 275 GHz. Above this spectrum is unallocated. There are various bands above 95 GHz in the United States and 86 GHz in the rest of the world that are allocated to fixed services, but there are no service rules in place, meaning that commercial high data rate services are prohibited. Nevertheless, there are research groups building prototypes at these high frequencies to demonstrate potential high data rate systems. For example, researchers in Japan have demonstrated 10-Gbps data transmission at 120 GHz one way over a 200-m distance using photonic techniques [16].

6.6   The Future?

Because the 10 GHz of spectrum allocated at 70 and 80 GHz is unprecedented in the history of spectrum management, the real potential for these bands may not yet even be envisioned. Nontraditional companies that think outside the box will be the innovators and instigators of many new and exciting applications in the future.

However, in the near term there are two areas where advances will be made in these frequency bands.

6.6.1   Increased Performance and Link Distance

Going forward, 70/80-GHz systems operating with higher modulation types will become available. One reason is to increase data throughput to much higher than the 1 Gbps widely available today. Prototype 70/80-GHz systems supporting 6-Gbps throughput have been demonstrated, and 64 QAM systems supporting up to 24 Gbps have been proposed [13, 14]. A second reason is to pass higher data rates in the narrower channel sizes being promoted in Europe. To pass a 1-Gbps gigabit Ethernet signal in a 250-MHz channel, a modulation of 64 or 128 QAM will be required, depending on implementation (see Table 6.3).

As modulation complexity increases, the signal-to-noise ratio and system linearity requirements increase, receiver sensitivity will fall and output power will drop because power amplifiers will need to be operated increasingly backed off from their saturated output power. Because of the resulting reduction in system gain, achievable link distances will fall.

For this reason, there will be a push to develop higher output power millimeter-wave devices. Currently the state of the art in 70/80 GHz power amplifiers is approximately 24 dBm (250 mW) output power at 1-dB compression. However, there is plenty of scope in the United States and European regulations for higher output powers. FCC Part 101 rules allow up to 35-dBm (3-W) output power (Table 6.1) and ETSI allows up to 30 dBm (1W) with certain antenna sizes, and up to 35 dBm for smaller antennas if ATPC is used (Figure 6.3). Although there are fabrication challenges in reaching these higher powers at the higher millimeter-wave frequencies, we will see advances in device performance, either as individual higher power MMIC devices or as packages of multiple devices combined in a multichip module to provide higher power capabilities.

It has also been proposed that adaptive antennas can be used for 70/80-GHz systems similar to those used for 60-GHz wireless LANs [14]. At the higher millimeter-wave frequencies, active arrays with integrated power amplifiers and antenna elements can provide coherent spatial power combining. Since multiple power amplifiers in parallel will have an output power proportional to the number of elements (less any combiner losses), and the effective aperture and antenna gain of a multiple antenna system is also proportional to the number of elements, the combination of both in an active system results in an n^2 improvement in possible EIRP. If an identical receive antenna array is utilized, the effective SNR may increase by n^3, depending on the level of signal correlation. Thus, a very high EIRP system may be achievable, with beam steering to support advanced applications or even just to preserve the basic problems of aligning and commissioning very high gain antennas. Regulatory EIRP limits of +55 dBW (300 kW) still have to be observed and place an ultimate limit on the size of such antenna arrays.

6.6.2   40 Gbps and 100 Gbps Systems

The real benefits of operating in the higher millimeter-wave bands are the wide spectrum channels that are available and the potential for very high data rate transmission that these wide bandwidths permit. In the wired world, work is underway to develop standards for 40-Gbps and 100-Gbps transmission. To support such data rates wirelessly, channel bandwidths of many tens of gigahertz will be required. Only at the higher upper-millimeter-wavelengths can such band-widths be achieved.

The two 5-GHz channels available at 70/80 GHz are not sufficient to support a 40-Gbps data rate using today's technologies. Even very high modulation schemes such as 256 QAM do not have the efficiency to compress such high data rates into even the widest 70/80-GHz channels. Therefore, a different approach is required if such data rates are to be supported in the future.

For this reason, there is interest in opening up spectrum at 140 GHz and 240 GHz for ultrahigh data rate applications. Similar to 94 GHz, both 140 GHz and 240 GHz correspond to atmospheric absorption windows. The molecular absorption peaks at 119 GHz, 183 GHz, and 325 GHz must be avoided, as they will limit transmission distances. However, there is sufficient spectrum to allocate 40 GHz of bandwidth from 125 to 165 GHz and 100 GHz from 200 to 300 GHz, as shown in Figure 6.16. With 40 GHz and 100 GHz of spectrum available, ultrahigh data rates of 40 Gbps and 100 Gbps could be supported. Although higher atmospheric attenuation occurs at these higher frequencies, rain fade is comparable to currently available 70/80-GHz systems (see Figure 2.10), meaning that practical links of up to about 1 km are quite conceivable.

Currently, fixed service operation in these higher frequency spectrum bands is prohibited, although individual links can be operated under temporary experimental licenses. There are numerous allocations above 100 GHz up to 275 GHz for fixed links, but the allocations are fragmented and relatively small

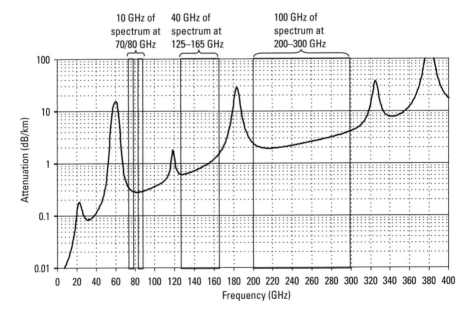

Figure 6.16 Possible spectrum allocations for ultrahigh capacity data rate transmissions.

and there are no service rules to govern their use. The remaining spectrum is allocated for radio astronomy, space research, and earth exploration.

To open up these proposed upper-millimeter-wave frequency bands, a motion must first be filed at the ITU's World Radio Conference (WRC) to modify the International Table of Allocations to reallocate the proposed 40-GHz and 100-GHz blocks of spectrum for fixed wireless use. Once the process to modify the International Table of Allocations is underway, a petition for rulemaking needs to be filed with the FCC to similarly modify the U.S. frequency allocation plan and to develop licensing and service rules for the bands. A similar petition would need to be filed with CEPT to initiate frequencies allocations in Europe, and then with ETSI to develop technical rules for the bands. History has shown that this process may take 10 years or more, from initiation to opening of the bands for commercial use.

6.7 Summary

There is much interest in the 70 GHz and higher millimeter-wave bands because of the wide spectrum channels that are available and the ability for systems to transmit very high data rates in these wide bandwidths. There is particular interest in the 71–76-GHz and 81–86-GHz E-band allocations because these bands are ITU-approved and have recently been made available in many of the major markets around the world. Although the technical rules for 70/80-GHz radios are very different in the United States and Europe, frequency allocations are consistent and commercial equipment is widely available that can deliver a 1-Gbps data rate while satisfying both regional requirements. The 70/80-GHz equipment will play an increasing role in cellular backhaul applications, particularly for shorter-distance, higher data rate requirements. In particular, the opportunity to replace fiber in connecting together remote radio heads for DAS applications is very exciting.

Architecturally, currently available 70/80-GHz systems are relatively simple. However, as more devices are adopted and applications mature, system sophistication will grow. The 70/80-GHz radios will be required to deliver higher data rates and to deliver it more spectrally efficiently. Systems will move away from the current two-level modulations to higher QPSK and QAM linear modulations, placing more stringent requirements on the system architecture and performance. There will be advances particularly in the modems, millimeter-wave circuitry, and antennas to deliver the increase performance characteristics to support higher modulations, power and data rates.

Finally, in order to move to ultrahigh data rates of 40 Gbps and even 100 Gbps, new spectrum allocations in the upper-millimeter-wave bands will be explored, opening another new frontier in wireless commercial services.

References

[1] Wells, J. A., "Multigigabit Wireless Technology at 70, 80 and 90 GHz," *RF Design*, 2006, pp. 50–58.

[2] FCC, Millimeter Wave 70-80-90 GHz Service, http://wireless.fcc.gov/services/index. htm?job=service_home&id=millimeter_wave.

[3] ECC Recommendation (05)07, "Radio Frequency Channel Arrangements for Fixed Service Systems Operating in the Bands 71-76 GHz and 81-86 GHz," 2009.

[4] RALI FX 20, "Millimetre Wave Point to Point (Self-Coordinated) Stations," 2007.

[5] PIB 22, "Fixed Service Bands in New Zealand," Issue 5, 2009.

[6] FCC, "Code of Federal Regulations, Title 47—Telecommunication, Part 101: Fixed Microwave Services," 2009.

[7] ETSI EN 302 217-3, "Fixed Radio Systems; Characteristics and Requirements for Point-to-Point Equipment and Antennas; Part 3: Equipment Operating in Frequency Bands Where Both Frequency Coordinated or Uncoordinated Deployment Might Be Applied; Harmonized EN Covering the Essential Requirements of Article 3.2 of the R&TTE Directive," version 1.3.1, 2009.

[8] CEPT/ERC/Recommendation 74-01E, "Unwanted Emissions in the Spurious Domain," 2005.

[9] ETSI EN 302 217-4-2, "Fixed Radio Systems; Characteristics and Requirements for Point-to-Point Equipment and Antennas; Part 4-2: Antennas; Harmonized EN Covering the Essential Requirements of Article 3.2 of the R&TTE Directive," version 1.5.1, 2010.

[10] PIB 38, "Radio Licence Certification Rules," Issue 6, 2009.

[11] Fontaine, L., "71–95 GHz Registration—'Light' Licensing with Interference Protection," *Mission Critical Communications,* Vol. 20, No. 8, 2005.

[12] OfW 369, "Guidance Notes for Self Co-Ordinated Licence and Interim Link Registration Process in the 71.125-75.875 GHz and 81.125-85.875GHz Bands," 2007.

[13] Dyadyuk V., et al., "A Multi-Gigabit Mm-Wave Communication System with Improved Spectral Efficiency," *IEEE Trans. on Microwave Theory and Techniques,* Vol. 55, No. 12, Part 2, 2007, pp. 2813–2821.

[14] Dyadyuk, V., J. D. Bunton, and Y. J. Guo, "Study on High Rate Long Range Wireless Communications in the 71–76 and 81–86 GHz Bands," *Proceedings of European Microwave Conference,* 2009, pp. 1314–1318.

[15] Fallica, M. T., "38, 60 & 82 GHz MMICs for High Capacity Communication Links," *Microwave Engineering Europe,* 2007.

[16] Hirata, A., et al., "120-GHz-Band Millimeter-Wave Photonic Wireless Link for 10-Gb/s Data Transmission," *IEEE Trans. Microwave Theory & Tech.,* Vol. 54, No. 5, 2006, pp. 1937–1044.

7

High Data Rate Wireless Link Design

7.1 Introduction

Chapter 2 discusses propagation characteristics, including free space loss, atmospheric attenuation, rain attenuation, and a variety of other mechanisms that limit radio transmissions. General high capacity wireless equipment is covered in Chapter 3, outlining various parameters that contribute towards the transmission of gigabit per second data across an air-path communication channel. Chapters 4, 5, and 6 detail specific characterizations for 6–40-GHz microwave, 60-GHz millimeter-wave, and 70/80-GHz and higher millimeter-wave radios, respectively. This chapter pulls all this information together to show how high capacity gigabit per second links can be designed and how such wireless systems can be characterized.

Two primary characteristics of high capacity wireless networks are link distance and system availability—the proportion of time that the link is operational and performing above a minimum performance threshold. The two are highly dependent and determined by a number of static and variable factors including equipment performance characteristics, location, and atmospheric conditions. The objective of a good link design is to build sufficient margin into the wireless system to overcome the variable parameters, while not overengineering the devices and causing unnecessary complexity or cost.

The primary channel variable for high capacity microwave and millimeter-wave wireless links is rain attenuation. There are two dominant statistical rain predication models—Crane and ITU—that allow rainfall rates to be forecast around the world. Both of these have strengths and weaknesses. A number of techniques are available that equipment providers and network architects can use to make commercial high data rate equipment more robust and resilient to the variable atmospheric conditions.

This chapter provides a detailed analysis of the unique factors that affect the link design for high capacity microwave and millimeter-wave radios. Typical commercial systems are analyzed and methods to predict and characterize their performance under real-life operating conditions are presented.

It should be noted that throughout this chapter the terms "availability" and "weather availability" are used interchangeably and both are used to refer to the strong dependence of high capacity wireless links on varying weather, and primarily high intensity rain. The concept of equipment availability, the amount of time a link is operational and not down due to equipment failure, is also an important concept for wireless links. The concepts of mean time between failures (MTBF) and mean time to repair (MTTR) are more general topics and applicable to all wireless devices. As such, they are not covered here.

7.2   Link Budget and Fade Margin

A primary characteristic for any wireless system is a link budget. A link budget is a method of accounting for all the gains and losses in a communications system, from the power out of the transmitter, through the free space wireless medium, to the received signal at the receiver. A link budget accounts for the attenuation of the transmitted signal due to propagation and atmospheric effects, as well as antenna gains and various implementation losses. At its most simplistic, a link budget is given by:

$$
\begin{aligned}
\text{Recieved Power (dBm)} &= \text{Transmitted Power (dBm)} \\
&+ \text{Gains (dB)} - \text{Losses (dB)}
\end{aligned}
\tag{7.1}
$$

To develop this further, it is useful to consider the signal path in a line-of-sight wireless system, as characterized by Figure 7.1. The electrical output power generated by the transmitter (Pout) is passed to an antenna for transmission. If the antenna is not directly mounted to the transmitter, a feeder loss associated with the coaxial cable or waveguide feed between the radio and the antenna needs to be accounted for. The antenna focuses the signal from the transmitter in a particular direction by an amount equivalent to the gain of the transmitter antenna (Gtx) to give the radiated power (EIRP) emitted from the device. As the signal then travels through free space, it undergoes rapid attenuation due to

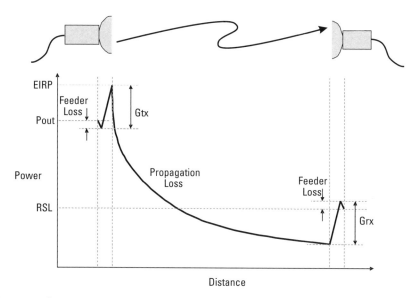

Figure 7.1 Conceptual representation of a link budget.

free space loss and atmospheric attenuation. As shown in Section 2.3, the free space loss (L) decreases the transmitted power level proportional to the inverse square of the distance from the transmitter and can be significant and usually dominates the link budget. At the opposite end of the link, the receive antenna collects the signal and concentrates it by an amount equivalent to the gain of the receive antenna (Grx). This is then passed to the receiver (possibly via a lossy feeder interface) for detection, further amplification, and processing. The power level at the receiver input is known as the receive signal level (RSL), and can be derived from knowledge of all these other variables and the link budget equation.

Thus, (7.1) can be more completely written as:

$$RSL \text{ (dBm)} = Pout \text{ (dBm)} + Gtx \text{ (dBi)} + Grx \text{ (dBi)}$$
$$-\text{feeder losses (dB)} - L \text{ (dB)} - \text{atmospheric attenuation (dB)} \quad (7.2)$$
$$-\text{implementation loss (dB)}$$

Note that an implementation loss has to be accounted for. Antennas are seldom perfectly aligned, and variations in the properties of the air path can add unaccounted for attenuation to wireless transmissions (for example, cross-polarization distortion and dispersion due to the wireless channel).

Equation (7.2) allows the receiver RSL to be determined, given a static, but realistic, wireless communication link. The purpose of a link budget is to ensure that the receiver RSL falls in the optimal region of the receiver's response curve. Such a response is shown in Figure 7.2, which is often known as a *bathtub curve*. If the RSL is too large, the receiver will overload, distorting the signal and increasing the bit error rate (BER) of the link. If the receiver overload is too great, the sensitive low noise amplifier can be permanently damaged. If the signal is too small, the receiver will have trouble extracting the wanted signal from the system noise, which will introduce significant bit errors into the system. For a correctly designed link, the RSL will be in the flat operating region of the receiver's response curve, where minimum bit errors occur in the link. Note that statistically no wireless link can operate with zero bit errors. Some background or residual BER will always occur. For a practical high capacity link, this level should be between 10^{-12} to 10^{-14}, depending on implementation, meaning that there will be approximately 100 to 10,000 bit errors per day in each direction for a 1-Gbps link.

The minimum signal that the receiver can detect—the receiver sensitivity—is usually quoted at a given performance level. This is typically at a system BER performance of 10^{-6}. For the receiver shown in Figure 7.2, the receive sensitivity at 10^{-6} BER is −70 dBm. The amount by which the receiver RSL exceeds the receiver sensitivity is called the fade margin, as depicted in Figure 7.3. This single parameter determines the margin built into the link to overcome variable changes in the wireless channel. These variations are known as *fades*. For an outdoor wireless link, the two principal causes of fades are rain and multipath. Multipath fading is dependent on various factors including frequency, path length, terrain, climate, and a number of other factors. Fortunately, for high data rate links, operating at higher frequencies over shorter distances with highly

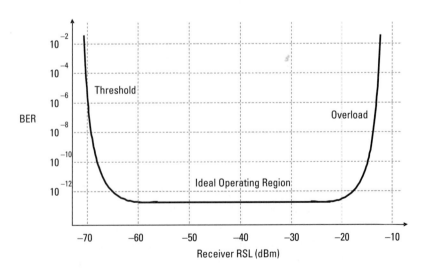

Figure 7.2 Typical receiver RSL curve.

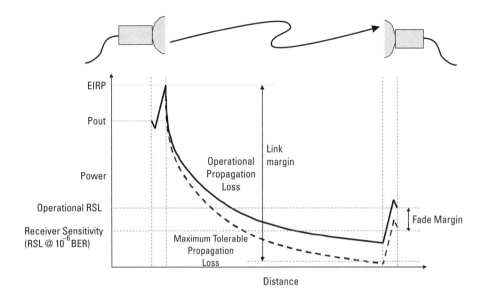

Figure 7.3 Conceptual representation of a fade margin.

directional antennas, multipath is generally not an issue, so it is not covered further here. Therefore, the principal purpose of a fade margin in a high capacity wireless link is to provide excess margin to overcome fades due to rain.

7.2.1   Example Link Budget

Consider a hypothetical high capacity link operating with the following real-world performance characteristics:

- Operational frequency band: 70/80 GHz;
- Operational frequencies: 73.5 GHz and 83.5 GHz FDD;
- Output power (Pout): 20 dBm;
- Receiver sensitivity (RSL at 10^{-6} BER): −65 dBm;
- Antenna gain (Gtx, Grx): 50 dBi (equivalent to approximately 2 ft (60 cm) diameter antennas);
- Feeder losses: 0 dB;
- Link distance: 2 km (1.25 miles).

Since this is an FDD link, the highest transmission frequency needs to be considered, since this will have the highest free space loss and hence the lowest fade margin (assuming all other radio parameters are equal). The link budget at 83.5 GHz is detailed in Table 7.1 and shows that a 2-km, 70/80-GHz link operating with the radio parameters above will have an RSL of −18 dBm under unfaded conditions and a fade margin of 47 dB.

In practice, this link budget suggests a well designed link. The fade margin is large at 47 dB and would allow fades equivalent to 23.5 dB/km before the system BER dropped to 10^{-6} or worse. Figure 2.11 shows that this is equivalent to rain falling at a rate of approximately 70 mm/hr (2.75 inches/hour). One area of concern with the link is that the unfaded receiver RSL of −18 dBm may be close to or within the overload region of the receiver RSL response (see Figure 7.2), and a higher-than-wanted background BER might be occurring. If it were, it would be sensible to reduce the transmitter output power a little to trade off the unfaded BER caused by small amounts of receiver overload with the fade margin providing resilience to rain fades.

7.2.2 Fade Margin Versus Link Distance

In the above link budget example, it is shown how a realistic 70/80-GHz link can operate over a 2-km (1.25-mile) distance. Clearly, the link can operate over longer distances, but with reduced fade margin and hence reduced resilience to overcome fading.

To determine the link's maximum operational distance, consider Figure 7.4, which shows attenuation versus distance for the example 83.5-GHz transmission previously considered. The attenuation here is the sum of the free space loss and the atmospheric attenuation. The free space loss dominates this attenuation.

Table 7.1
Link Budget for Hypothetical 70/80-GHz High Capacity System

Parameter	Value	Cumulative Power Level	Notes
Transmitter output power (Pout)	20 dBm	20 dBm	
Less transmitter feeder loss	0 dB	20 dBm	
Plus transmitter antenna gain (Gtx)	50 dBi	70 dBm	EIRP with ~ 2-ft (60-cm) antenna
Less free space loss and atmospheric attenuation			
Free space loss (L)	137 dB		From (2.11)
Atmospheric attenuation	1 dB		From Table 2.2, ~ 0.5 dB/km
Total	138 dB	−68 dBm	Power at receiver antenna input
Plus receiver antenna gain (Grx)	50 dBi	−18 dBm	
Less receiver feeder loss	0 dB	−18 dBm	Unfaded receiver RSL
Less receiver sensitivity (RSL at 10^{-6} BER)	−65 dBm	47 dB	Fade Margin

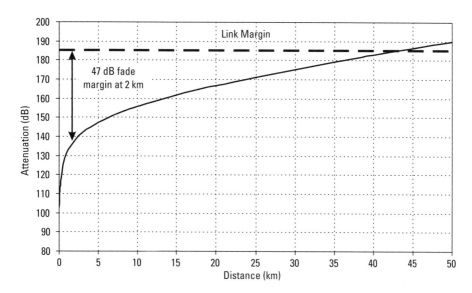

Figure 7.4 Propagation attenuation versus distance and link margin for a hypothetical 70/80-GHz high capacity system.

Also shown is the radio's link margin, a parameter commonly quoted on equipment data sheets. As shown in Figure 7.3, link margin is the difference between the radiated transmitter power (the EIRP) and the signal level at the receiver antenna when the receiver is operating at the sensitivity level. This is therefore the link operating parameters for a 0-dB fade margin and is equivalent to the maximum amount of propagation loss that a link can tolerate. In the previous example, the 70/80-GHz link has a link margin of 185 dB (determined by the difference between the EIRP (+70 dBm) and the signal at the receive antenna when the link is operating at the sensitivity level (–65 dBm – 50 dBi = –115 dBm)). From Figure 7.4, a perfectly installed 70/80-GHz link with a 185-dB link margin can in theory operate over a 42-km (26-mile) distance.

Clearly, this is unrealistic. First, the link was assumed to be perfectly aligned. Second, the analysis considers only the unfaded case. Any antenna misalignments, variations in channel characteristics or moisture in the atmosphere will cause such a long transmission to fail.

Since rain is the primary fading mechanism for outdoor high capacity wireless links, consider now Figure 7.5, which considers a much more realistic scenario. Here the rain faded propagation loss at 83.5 GHz has been plotted for a variety of rain scenarios. In addition, the link margin has been reduced by 5 dB to account for real-world implementation losses (2 dB of pointing errors per antenna and 1 dB of channel impairment has been assumed). Now it can be seen that to overcome real life rain fades, practical 70/80-GHz links are limited to about a 5-km (approximately 3-mile) distance depending on the requirements for the link's robustness to rain.

Figure 7.5 is also useful for determining the effect of device performance parameters on link distance. In the earlier example, 2-ft (60-cm) diameter antennas with 50-dBi gain were used. In the United States, the minimum antenna size is 43 dBi, equivalent to about a 1-ft (30-cm) antenna. If the link analyzed was to use these smaller antennas for both transmit and receive, a link margin reduction of 14 dB would result, yielding a new practical link margin of 166 dB. From Figure 7.5 it can be seen that halving the antenna sizes reduces the effective link distance by about one-third. (For example, a link able to withstand heavy rain falling at a 25 mm/hr rate would be reduced from 3.2 km to about 2.3 km.) Alternatively, the effect of increased output power can be seen. In the United States and Europe, the maximum 70/80-GHz EIRP permitted is +85 dBm, or 15 dB higher than in the earlier example. Considering now a 15-dB higher link margin value of 195 dB, it can be seen that about a one-third increase in link distance can be achieved.

A similar analysis can be done for both a high capacity microwave radio and a 60-GHz millimeter-wave radio designed for outdoor use to compare how they perform over distance. A 23-GHz microwave radio providing 1.4 Gbps via dual channel dual polarization using a 256 QAM modulation is assumed. The following parameters were used, which are typical for commercial high capacity 23-GHz radio products:

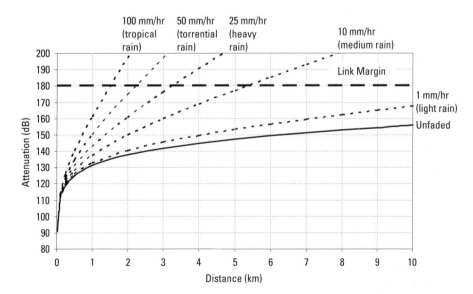

Figure 7.5 Rain faded propagation attenuation versus distance and link margin for a realistic 70/80-GHz high capacity system.

- Frequency: 23 GHz;
- Output power (Pout): 20 dBm;
- Receiver sensitivity (RSL at 10^{-6} BER): –60 dBm;
- 2-ft (60-cm) antenna gain (Gtx, Grx): 40 dBi;
- Feeder losses: 3 dB per antenna (for adjacent channel combiner and cross-polarization filter);
- Implementation loss: 3 dB.

For this radio, implementation loss has been reduced to 3 dB, since the antennas are easier to align and any cross-polarization distortion in the channel will be compensated for by the XPIC circuitry. Thus, this typical 23-GHz high capacity radio will have a link margin of 151 dB. This is a lot less than a 70/80-GHz radio because of the much smaller antenna gain at the lower transmission frequency. Also, the added feeder losses are a result of the combiner circuitry to allow cross-polarized and multichannel operations. In addition, the relatively low transmitter power and limited receiver sensitivity are a result of the stringent system requirements for a 256 QAM operation. A lower capacity sub-1 Gbps 23-GHz link will have a much improved link margin.

For the 60-GHz outdoor radio, it is assumed that the radio provides 1.0 Gbps and is operating at the FCC EIRP limit, with all other parameters typical for commercial two-level modulation systems:

- EIRP: 40 dBm (the FCC limit);
- Receiver sensitivity (RSL at 10^{-6} BER): –65 dBm;
- 2-ft (60-cm) antenna gain (Gtx, Grx): 48 dBi;
- Feeder losses: 0 dB;
- Implementation loss: 5 dB.

Here a link margin of 148 dB is calculated. Note that a 2-ft (60-cm) antenna was used for all the above examples to allow proper comparisons to be made. Larger aperture (and hence higher gain) antennas are readily available at 23 GHz and other microwave frequencies, and so link margins can be increased for the microwave products considered here.

Figures 7.6 and 7.7 show the rain faded propagation and link margin characteristics for the high capacity 23-GHz and 60-GHz link, respectively. For the 23-GHz radio, it can be seen that despite the reduced link margin, relatively long distances can still be achieved. Useable links can be achieved up to about 8 km (5 miles). For the 60-GHz radio, however, the limitations of the very high atmospheric attenuation can be seen. It is not possible for any 60-GHz link to

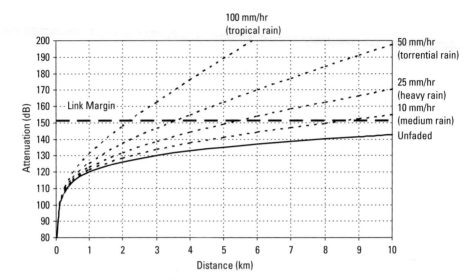

Figure 7.6 Rain faded propagation attenuation versus distance and link margin for a realistic 23-GHz gigabit-per-second system.

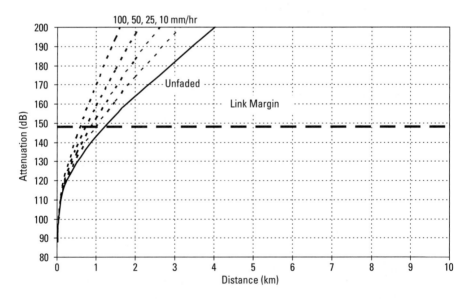

Figure 7.7 Rain faded propagation attenuation versus distance and link margin for a realistic 60-GHz outdoor gigabit-per-second system.

realistically operate beyond 1 km (0.6 mile), even when operating at the maximum regulatory limits.

7.3   Link Availability and Distance

It has already been shown that rain causes attenuation of radio wave transmissions above about 10 GHz, and that wireless links need to be designed and installed with a fade margin to provide protection against such rain fades. A question facing every link designer is how much fade margin should be engineered into each link. If too big a fade margin is chosen, the link will be overengineered and the cost spent on large antennas and higher power transmitters may be wasteful. If too low a fade margin is chosen, the link will experience more outages than perhaps can be tolerated. The correct fade margin is therefore a tradeoff between technical performance, cost, and desired link performance. This tradeoff is often quantified in terms of link availability.

7.3.1   Availability

Since outages caused by rain fades are relatively long (of order minutes and longer), it is usual to quote link performances in terms of weather availability—the statistical amount of time that the link is operating with better than a given performance—when subjected to rain fades. If the fade exceeds the link's fade margin, an outage will occur and the link is said to be unavailable. If a link experiences outages 1% of the time, it is said to have an availability of 99%.

A widely used reference for high capacity link availability is 99.999%, often known as "five-nines." This means that the link would experience outages for 0.001% of the time, or about 5 minutes a year. This is often known as carrier-class availability, because it has long been the telecom operators' benchmark for wired and wireless telecommunications. Another widely used metric is 99.9% availability. This is known as enterprise-class, as it is a reduced standard for the less demanding consumer/commercial market. A link with 99.9% availability would experience 0.1% outages, or be unavailable for about 8 hours per year. A summary of availability and outage times is given in Table 7.2.

To be able to specify a particular link availability, an understanding of the outage mechanisms are required. Since the primary cause of outages in a high capacity link is rain, a good understanding of rain statistics is required.

7.3.2   Rain Statistics

A number of statistical rain attenuation models have been developed to allow link availability metrics to be established. Some models are based on rain rate and path attenuation measurements at particular locations, which are then extrapolated to wider areas. Others are based on physical models that use statistical information on rain rates, with scattering models using rain drop size, shape, orientation, and temperature information.

Table 7.2
Common Availability and Outage Conversions

Availability	Outage Per Year	Average Outage Per Month	Average Outage Per Day	Nomenclature
99%	88 hours	7.3 hours	14 minutes	
99.9%	8.8 hours	44 minutes	1.4 minutes	Three-nines, enterprise-class
99.99%	53 minutes	4.4 minutes	8.6 seconds	
99.999%	5.3 minutes	26 seconds	0.86 second	Five-nines, carrier-class
99.9999%	32 seconds	2.6 seconds	0.086 second	

The two most widely used are the Crane model and the ITU model. The Crane model is widely used in the United States, with the ITU method more commonly accepted in the rest of the world. It is important to note that both models predict rain intensities—the rate of rainfall. The amount of rain is of no value in link availability modeling. It is not how much it rains; it is how hard it rains and how often.

7.3.2.1   Crane Rain Model

In 1980, Dr. Robert Crane [1] established a global map of eight rain climate zones with similar rainfall statistics. This data is based on several decades of rain-rate measurements, arranged into similar zones taking account of rain types, thunderstorm activity, and climate maps. For each of these rain zones, a rain-rate distribution was determined based on average measurements in that zone.

Over the years, the Crane model has been refined a number of times, yielding the Crane Global model, the Crane two-component model, and the revised Crane two-component model, all of which produce slightly different (and arguably increasingly more accurate) estimates of the long-term mean rain fade probability. In the most recent model, there are eight rain climate zones that cover North America. These are shown in Figure 7.8. The frequency of different rainfall events in each of these North American rain climate zones is given in Table 7.3.

7.3.2.2   ITU Rain Model

The ITU-R Study Group 3 [formerly part of the CCIR (Consultative Committee on International Radio)] has also established a rain climate zone model for the prediction of rain-rate distributions [2]. In its earlier versions, the ITU model also defined a number of regional zones with similar precipitation characteristics. However, unlike the eight zones in the earlier Crane model, the ITU model defined 15 rain climate zones, as shown in Figure 7.9. The frequency for different rainfall events in each of these rain climate zones is given in Table 7.4.

Figure 7.8   Crane rain climate regions for North America.

Table 7.3
Crane Model Rainfall Rate Versus Percentage of Year That
Rainfall Rate Is Exceeded in Each North American Rain Region

	Rain Region and Rain Rate (mm/hr)							
% of year	B1	B2	C	D1	D2	D3	E	F
0.001%	45	70	78	90	108	126	165	66
0.002%	34	54	62	72	89	106	144	51
0.005%	22	35	41	50	64.5	80.5	118	34
0.01%	15.5	23.5	28	35.5	49	63	98	23
0.02%	11	16	18	24	35	48	78	15
0.05%	6.4	9.5	11	14.5	22	32	52	8.3
0.1%	4.2	6.1	7.2	9.8	14.5	22	35	5.2
0.2%	2.8	4	4.8	6.4	9.5	14.5	21	3.1
0.5%	1.5	2.3	2.7	3.6	5.2	7.8	20.6	1.4
1%	1	1.5	1.8	2.2	3	4.7	6	0.7
2%	0.5	0.8	1.1	1.2	1.5	1.9	2.9	0.2
5%	0.2	0.3	0.5	0	0	0.9	0.5	0

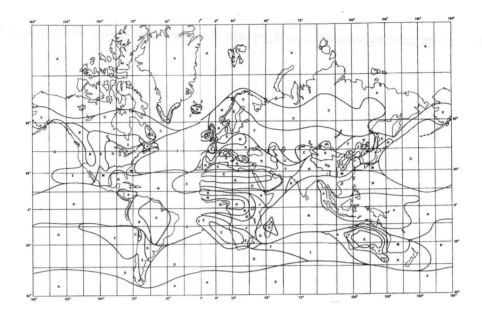

Figure 7.9 ITU rain climate zones.

Table 7.4
Rainfall Rate Versus Percentage of Year That Rainfall Rate Is Exceeded in Each ITU Rain Region

% of year	Rain Region and Rain Rate (mm/hr)														
	A	B	C	D	E	F	G	H	J	K	L	M	N	P	Q
1%	<0.1	0.5	0.7	2.1	0.6	0.7	3	2	8	1.5	2	4	5	2	24
0.3%	0.8	2	2.8	4.5	2.4	4.5	7	4	13	4.2	7	11	15	34	49
0.1%	2	3	5	8	6	8	12	10	20	12	15	22	35	65	72
0.03%	5	6	9	13	12	15	20	18	28	23	33	40	65	105	96
0.01%	8	12	15	19	22	28	30	32	35	42	60	63	95	145	115
0.003%	14	21	26	29	41	54	45	55	45	70	105	95	140	200	142
0.001%	22	32	42	42	70	78	65	83	55	100	150	120	180	250	170

   In 1999, the ITU changed methodology to improve the resolution of their rain model. This model replaces the 15 rain bands and the seven availability probabilities with a computational approach that provides rainfall model parameters for a wider variety of probabilities based on more localized geographical coordinates. Data is provide based on location, with a latitudinal resolution from +90°N to –90°S in 1.5° steps, and a longitudinal resolution from 0° to

360° in 1.5° steps, yielding 28,800 individual data points. Rainfall information was derived from 15 years of data from the European Centre of Medium-Range Weather Forecasting. From this, a more localized prediction of rainfall for a particular link location can be made.

In order to make a localized rainfall estimate, the ITU-R Study Group 3 provides a MATLAB implementation of its model and a number of associated data files containing relevant input parameters. First, three variables (P_{r6}, M_c, and M_s) are extracted for the four closest latitude and longitude points on the 1.5° geographical grid surrounding the desired link coordinates. Then a bilinear interpolation is performed to obtain the values $P_{r6}(\text{Lat,Lon})$, $M_c(\text{Lat,Lon})$, and $M_s(\text{Lat,Lon})$ at the desired location. From this, the probability of rain, P_0, is derived:

$$P_0\left(\text{Lat,Lon}\right) = P_{r6}\left(\text{Lat,Lon}\right)\left(1 - e^{-0.0117\left(M_S\left(\text{Lat,Lon}\right)/P_{r6}\left(\text{Lat,Lon}\right)\right)}\right) \tag{7.3}$$

If the result of this operation is undetermined (not a number), the probability of rain $P_0(\text{Lat,Lon})$ is equal to zero and consequently the rainfall intensity is also zero. For all other cases, the rainfall rate, R_p, exceeded for p % of an average year can be determined from:

$$R_p\left(\text{Lat,Lon}\right) = \frac{-B + \sqrt{B^2 - 4AC}}{2A} \tag{7.4}$$

where:

$A = a\, b$

$B = a + \text{dn}(p/P_0(\text{Lat,Lon}))$

$C = \ln(\ p/P_0(\text{Lat,Lon}))$

$a = 1.11$

$b = \dfrac{\left(M_c\left(\text{Lat,Lon}\right) + M_S\left(\text{Lat,Lon}\right)\right)}{22932 P_0}$

$c = 31.5 b$

Figure 7.10 shows the global rainfall levels exceeded for 0.001% of the year using this methodology. The intense rainfall along the equatorial zones can be clearly seen, as can the dry African deserts and arid Middle Eastern and polar areas.

Both the Crane and the ITU models are widely used, although the Crane model tends to be favored in the United States and the ITU model in the rest of the world. It is not clear whether one model is more accurate than the other, and a number of comparisons have been made between the two. For example, one

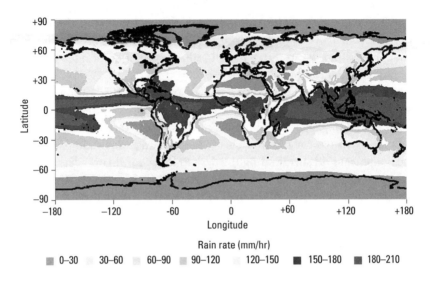

Figure 7.10 Global rainfall rates exceeded for 0.001% of the year using latest ITU rain model.

analysis [3] concludes that the various Crane models tend to produce higher rain fades than the ITU model. However, the year-to-year and location-to-location uncertainty of all the models are relatively large. Table 7.5 shows the various rainfall rates predicated by the three models for various major cities around the world. In most cases rough agreement can be seen between the three models. Mexico City falls close to a regional boundary on both the Crane and ITU zone models, so the new ITU model which allows interpolation presumably provides a better estimate. For Riyadh, a large discrepancy between the Crane model and both ITU models is shown. This is likely due to the Crane model having fewer rain zones and so not allowing localized climates (for example, deserts and microclimates) to be properly modeled.

Table 7.5
Rainfall Rate (in mm/hr) Versus Percentage of Year That Rainfall Rate Is Exceeded in Six Major Cities Using the Crane and Both ITU Rain Models

City	Crane Model			Old ITU Zone Model			New ITU Discrete Model		
	Zone	0.1%	0.001%	Zone	0.1%	0.001%	Lat., Long.	0.1%	0.001%
London	C	7.2	78	E	6	70	51.5, −0.17	7.1	62
Mexico City	D	14.5	108	M	22	120	19.42, −99.17	7.3	72
New York	D	14.5	108	K	12	100	40.75, −73.98	11.8	100
Riyadh	F	5.2	66	A	2	22	24.41, 46.42	1.3	21
Singapore	H	51	251	P	65	250	1.28, 103.85	48.1	177
Sydney	D	14.5	108	M	22	120	−33.92, 151.17	11.3	103

In the author's opinion, the new ITU discrete model is preferred. Rainfall rate is a smooth function of position unlike the step functions caused by discrete rain regions with other models. Rainfall rate is a function of percentage probability of rain, so calculations based on arbitrary percentages are possible. Finally, since the ITU model is a widely adopted ITU recommendation that forms the consensus among a group of international experts, it is believed to be the best method to estimate fade events.

Despite this recommendation, if local measured rainfall intensity data is available, that should always be considered before any model. A variety of hydrometers are available for such a purpose, such as a tipping bucket or a drop counter. A small integration time, such as 1 minute, should be used for such local measurements.

7.3.3 Rain Attenuation

In order to determine usable link distances, the attenuation due to rain at the desired availability level needs to be known. Section 2.5 details attenuation due to rain and shows that rain attenuation can be modeled as a function of frequency, rain rate, polarization, and slant angle. Using this methodology, the rain attenuation for the six cities detailed in Table 7.5 at various frequencies can be determined. This is shown in Table 7.6. Note that this analysis uses the new ITU rain model data and considers just vertical polarized links and vertically falling rain.

As an example of how to use this data, consider a wireless link in New York City. For this link to operate with 99.999% weather availability (5 minutes of outages per year), one would need to know the outage conditions for 0.001% of a year. Per Table 7.6, this is when the rain rate exceeds 100 mm/hour (4.0 inch/

Table 7.6
Rainfall Attenuation Versus Percentage of Year That Attenuation Is Exceeded for Six Major Cities

City	Rain Intensity (mm/hr)		Rain Attenuation (dB/km)					
			23 GHz		60 GHz		70/80 GHz*	
	0.1%	0.001%	0.1%	0.001%	0.1%	0.001%	0.1%	0.001%
London	7.1	62	0.85	6.8	3.7	18.7	4.8	21.4
Mexico City	7.3	72	0.87	7.9	3.8	20.9	4.9	23.8
New York	11.8	100	1.4	10.8	5.4	26.8	6.8	29.9
Riyadh	1.3	21	0.17	2.4	1.0	8.3	1.5	10.1
Singapore	48.1	177	5.4	18.8	15.5	41.0	17.9	44.5
Sydney	11.3	103	1.33	11.1	5.2	27.4	6.6	30.6

* Rain attenuation calculated at 83.5 GHz.

hour). Thus to achieve 99.999% weather availability, any wireless link in New York City would need to be designed with a fade margin sufficient to overcome 100 mm/hour rain intensity. If the link was operating at 70/80 GHz, it would have to have a fade margin sufficient to overcome about 30-dB/km rain attenuation, in addition to the other attenuation parameters in the link budget equation (free space loss and atmospheric attenuation).

Now suppose an identical 99.999% availability link was to be installed in the dry conditions of Riyadh, Saudi Arabia. Now the link has to overcome only 21 mm/hr rain, equivalent to approximately 10-dB/km rain attenuation. Given the much lower potential attenuation, a significantly longer link can be established. Alternatively, for a similar transmission distance link, a system operational in Riyadh would have a much higher weather availability (and hence much greater uptime) than a similar distance link in New York.

7.3.4 Rain Cells

Given that for higher frequency bands, rain attenuation can be significant; knowledge of the characteristics of rain cells is useful. This enables localized rain events to be modeled, since it unlikely that intense rain is falling across the full linear length of the radio path.

With a rain cell model, only the portion of the radio path within the rain cell experiences full rain attenuation. The region outside the rain cell may still be attenuated, but at a lower level than in the rain cell (since it will probably still be raining, but at a much lighter intensity level than inside the rain cell). Figure 7.11 shows this scenario.

One model for rain cells, their rainfall characteristics and resulting attenuation is described by [4]. Here the typical diameter of a rain cell d_c (in kilometers) is modeled as a function of the rainfall rate by:

$$d_c = 3.3R^{-0.08} \tag{7.5}$$

where R is the rainfall rate in mm/hr. Figure 7.12 illustrates the diameter of a rain cell as a function of rainfall rate as predicted by this model. It can be seen

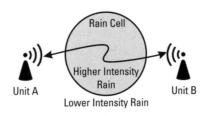

Figure 7.11 Rain cell model.

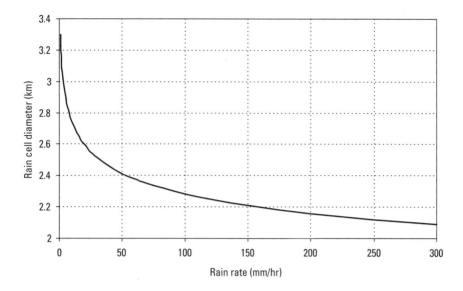

Figure 7.12 Diameter of a rain cell.

that light intensity rain cells are typically around 3 km in diameter. As intensity of rain increases, cell sizes can decrease by up to 50%, but still remain larger than 2 km.

Within the rain cell, an equal rain rate is assumed, and the resulting attenuation can be determined as per previous techniques [see (2.14)]. The specific attenuation in the area outside the rain cell decreases as the distance from the center of the rain cell increases, since it will still be raining, but at a steadily reducing intensity. The following characterization for the attenuation Γ (in dB/km) at a distance d (in kilometers) from the center of the rain cell can be used [4]:

$$\Gamma = \begin{cases} \gamma_R & \text{for } d \leq \frac{d_c}{2} \\\\ \gamma_R \cdot \dfrac{\exp\left\{\exp\left(-\dfrac{d - \frac{d_c}{2}}{r_m}\right)\right\}}{\cos(\varepsilon)} & \text{for } d > \frac{d_c}{2} \end{cases} \tag{7.6}$$

where γ_R is the specific attenuation for the rain rate R, ε is the elevation angle, and r_m is the scale length for rain attenuation, given by:

$$r_m = 600R^{-0.5}10^{-(R+1)^{0.19}} \tag{7.7}$$

Figure 7.13 illustrates the typical specific rain attenuation versus distance from the center of a rain cell for a 70/80-GHz link under various rain scenarios.

This analysis shows that for longer-distance links, particularly at the higher transmission frequencies, it is not realistic to apply the highest levels of attenuation associated with intense rain rates across the full length of the link. However, in traditional link profile models, this is the norm, suggesting that derived distances and availabilities would usually be conservative. In practice, the high attenuation values should only be applied to the section of the path that is experiencing the heavy rain. Typically, this section is only 2 km of the total path, so the attenuation should be modeled accordingly.

7.3.5   Example High Capacity Link Availabilities and Distances

To illustrate some real life path profile link distances and availability performances, consider the three hypothetical high capacity radios considered earlier, all carrying 1 Gbps or greater data rate for outdoor wireless transmissions. Table 7.7 summarizes the technical parameters for each of these systems. All parameters are typical for commercially available gigabit-per-second equipment. The 23-GHz unit is a 256 QAM dual channel cross-polarization microwave radio providing 1.4 Gbps in two 56-MHz channels using cross-polarized transmissions. The 60-GHz and 70/80-GHz radios deliver 1 Gbps using much lower modulation schemes and wider occupied bandwidths.

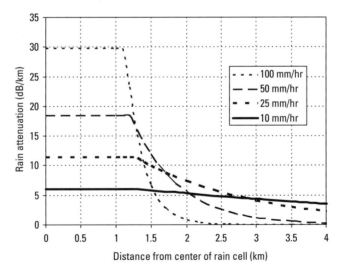

Figure 7.13 Rain rate attenuation versus distance from center of a typical rain cell for a 70/80-GHz link (calculations performed at 83.5 GHz).

Table 7.7
Typical Operations Parameters for a Range of High Capacity Radio Systems

Parameter	Typical High Capacity Radio Operational Values		
Frequency	23 GHz	60 GHz	70/80 GHz
Data Rate	1.4 Gbps	1.0 Gbps	1.0 Gbps
Transmission	Dual-channel cross-polarization	Single-channel single polarization	Single-channel single polarization
Operational parameters	256 QAM, 2 × 56 MHz channels with XPIC	BPSK, 1.5-GHz occupied bandwidth	BPSK, 1.5-GHz occupied bandwidth
Transmitter			
Pout	20 dBm		20 dBm
Feeder Loss	3 dB		0 dB
Antenna Gain [a]	40 dBi		50 dBi
EIRP		40 dBm [b]	
Receiver			
RSL at 10^{-6} BER	−60 dBm	−65 dBm	−65 dBm
Feeder Loss	3 dB	0 dBm	0 dBm
Antenna Gain [a]	40 dBi	48 dBi	50 dBi
Implementation Loss	3 dB	5 dB	5 dB
Link Margin	151 dB	148 dB	180 dB

[a] 2 ft (60 cm) antennas assumed for all frequencies; [b] FCC maximum EIRP limit.

In order to determine link profile characteristics, we need to solve the link budget equation using the operational parameters for the wireless link being analyzed. When the fade margin is zero, the link budget equation (7.2) can be rewritten as:

$$
\begin{aligned}
\text{Link Margin} = &\ L(\text{freq, distance}) \\
&+ \text{atmospheric loss(freq, lat, distance)} \\
&+ \text{rain attenuation(lat, long, freq, distance, availability)}
\end{aligned}
\tag{7.8}
$$

It can be seen that (7.8) is a function of numerous variables. Solving using the typical radio characteristics in Table 7.7, one can arrive at the distance and availability tradeoffs for the link. Table 7.8 shows the distances that can be achieved for both 99.9% and 99.999% weather availability for the high capacity 23-GHz, 60-GHz, and 70/80-GHz links for several major cities. Figures 7.14

Table 7.8
Link Distances Versus Weather Availability for Six Major Cities

City	Link Distance (km)					
	23 GHz		60 GHz		70/80 GHz*	
	99.9%	99.999%	99.9%	99.999%	99.9%	99.999%
London	11.5	3.1	1.09	0.70	6.8	2.0
Mexico City	9.2	2.7	1.09	0.67	6.2	1.8
New York	8.6	2.2	0.90	0.56	5.1	1.5
Riyadh	19.9	6.3	1.05	0.83	15.2	3.7
Singapore	3.5	1.5	0.76	0.56	2.3	1.1
Sydney	8.8	2.2	0.91	0.56	5.2	1.5

* Calculations performed at 83.5 GHz.

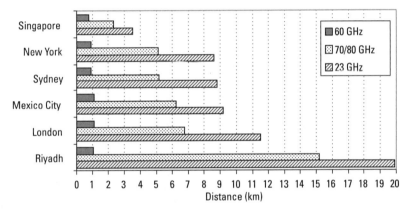

Figure 7.14 Link distances at 99.9% weather availability for typical high capacity 23-GHz, 60-GHz, and 70/80-GHz links.

and 7.15 show the same information diagrammatically, for 99.9% and 99.999% weather availability, respectively.

From this analysis, it can be seen that 60 GHz is a short-range technology, reaching typically 0.5 to 0.75 km for high carrier-class availability, and about 1 km for enterprise-class 99.9% availability, even when operated at the maximum FCC limits and large 2-ft (60-cm) antennas. The 70/80 GHz is a mid-range technology. Excluding the wet and dry climate extremes, 70/80 GHz offers about 1.5- to 2-km transmissions for 99.999% weather availability, and 5 to 7 km for 99.9%. Microwave radios provide the longest-distance high capacity wireless links. The 23-GHz example considered offers about a 50% premium on the 70/80-GHz radios—typically 2 to 3 km for carrier-class availability and about 10 km for lesser 99.9% weather availability.

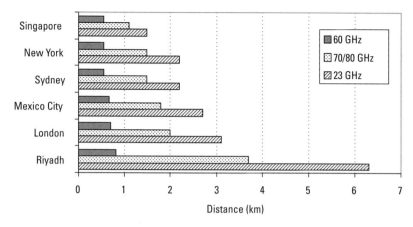

Figure 7.15 Link distances at 99.999% weather availability for typical high capacity 23-GHz, 60-GHz, and 70/80-GHz links.

Longer link distances can be achieved by increasing operational parameters, such as output power and antenna size. Also, by relaxing availability requirements and allowing more rain outages, longer link distances can be engineered. Finally, by dropping the data rate from the 1-Gbps and higher data rates considered here to values in the hundreds of megabits per second, more robust system parameters and hence longer-distance transmissions or higher availability numbers can be achieved.

7.4 Improving High Capacity Link Availability and Distance

In addition to increasing output power and antenna sizes, there are a number of additional techniques that can be used to increase the availability of a high data rate wireless link or network. This can result in either a more robust link or network or can allow link distances to be increased for a given availability requirement.

7.4.1 Automatic Transmit Power Control

Automatic transmit power control (ATPC) is a technique widely used in microwave radios whereby the transmitter power is deliberately reduced during clear air, nonfading conditions to reduce the risk of frequency interference to nearby wireless devices. During a rain fade, the system automatically increases power back up to the maximum level, restoring the full fade margin and allowing the link to overcome the effects or rain or other impairments. This is done by monitoring the far-end receiver level and feeding this information back to the transmitter.

ATPC has not been widely adopted in high capacity millimeter-wave systems. This is surprising since without ATPC, it is difficult for short- and medium-range millimeter-wave links to achieve high weather availabilities.

Consider a high capacity millimeter-wave link that does not employ ATPC, with a receiver characteristic as shown in Figure 7.2. For short and medium distance links, the transmitter output power often needs to be set at a level below its maximum value to ensure the received signal does not overload the receiver during clear air conditions. Therefore, knowledge of the link distance is required prior to setting the transmitter level, and manual tuning is required during link setup and commissioning to set this transmitter power. Since the link is operating at less than its specified output power, its fade margin is reduced and its ability to overcome rain fades is diminished.

In an ATPC-enabled link, no knowledge of the link distance is required. During clear air conditions, the transmitter will automatically set its output power to a level low enough to provide a good quality signal to the receiver, but well away from the overload point. During a rain fade, the drop in receiver level will be detected and the transmitter power will automatically increase to maintain a good quality signal at the receiver. By automatically altering the transmit level, the receiver will stay within a safe operating region, reducing the risk of bit errors. For the deepest fades, the transmitter will increase to its maximum value, and offer the full fade margin to provide resilience against the rain fade. As such, both the dynamic range of the receiver and the transmitter are available to provide margin against severe rain attenuation.

Not only does ATPC make a link more robust to rain fades, it is also an important feature for dense radio deployments. In a dense ATPC-enabled radio network, all transmitters will be operating at a backed-off power level during clear air conditions, meaning that radio emissions and hence the risk of potential interferences will be low. Although the power levels will increase during rain, the additional attenuation of the rain will still minimize any potential interference sources. Thus, ATPC-enabled radios can be much more densely deployed than non-ATPC equipment.

ATPC is so important for dense network designs that it is mandatory in some countries and some frequency bands. Although both the FCC and ETSI considered making ATPC mandatory for the millimeter-wave bands, it is currently an optional feature.

7.4.2   Adaptive Coding and Modulation

Adaptive coding and modulation (ACM) is a technique that has recently become popular for equipment operating in the 6–40-GHz microwave bands to combat rain fades and provide improvements in link availabilities. ACM enables the wireless system to adapt to link conditions, by changing modulation, power, and perhaps FEC coding to trade off data throughput with operational performance at any point in time. This technique was originally developed in lower-frequency license-exempt radios as a way to combat interference and multipath reflections.

As shown earlier, high capacity wireless links are traditionally engineered to ensure a target availability percentage, for example, 99.999%. In order to achieve this, system parameters such as antenna size, channel size, output power, and modulation are chosen to allow a sufficient fade margin to overcome the worst-case atmospheric conditions that only occur only 0.001% of the time. If the high capacity data traffic is such that the need for high capacity throughput can be relaxed for short periods of time during the worst outages, then adaptive modulation can be used to optimize the operational performance of the radio.

Figure 7.16 shows how ACM can be implemented on a high capacity microwave radio. To achieve maximum throughput, 256-QAM modulation is employed. Traditionally, for 99.999% weather availability, the link would require a large fade margin, which means either large antennas or high output power would be required, or the link distance would be limited to achieve this fade margin. For an ACM-enable link designed to meet 99.999% availability, the

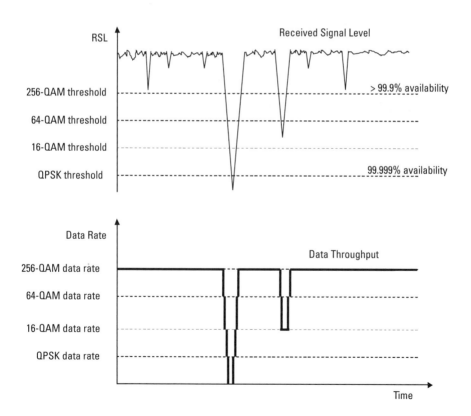

Figure 7.16 Effect of ACM implementation on data throughput.

fade margin at 256 QAM is set at a much reduced level, for example, 99.95%. This means that smaller antennas or lower output power can be used, or the link can be stretched over a longer distance. For all but 0.05% of the year (4 hours), the link would be operating with 256-QAM performance and maximum data throughput. During the 4 hours of fading, the link will progressively decrease the modulation to provide robustness against the rain fade, but at the expense of reduced data rate. If the lowest modulation supported is engineered at the 99.999% availability fade margin, the link will still operate at 99.999% availability. However, by trading off the throughput during the short time that rain fades occur, the overall operating parameters of the radio can be optimized.

ACM is not applicable for all types of data traffic. For traditional TDM traffic, where a strict fixed data throughput needs to be maintained, ACM cannot be used. However, for Ethernet traffic, where the payload data rate varies and links are often set up with oversubscription and quality of service to make sure that the highest priority traffic is transmitted, ACM is very appropriate.

Although ACM has been widely embraced by the microwave radio community, with many system providers offering ACM-enabled high capacity wireless systems, the network operators are slow to adopt the technology. The network-level implications of coping with unpredictable and perhaps rapidly varying data rates are still not established. Early trials have shown many configuration problems with nonwireless network peripherals, which have resulted in a number of networks designed with ACM being reengineered for fixed data rates.

7.4.3   Adaptive Rate

ACM is not applicable to all radio types. Current commercially available 60-GHz and 70/80-GHz radios only operate with low-order modulations, taking advantaged of the large channel sizes available to transmit 1-Gbps and higher data rates. As such, ACM is of limited use because there is very often no lower order modulation to fall back to. For this reason, some millimeter-wave systems employ a technique called adaptive rate, whereby the modulation is kept constant, but the data throughput is reduced during rain fades. By reducing the data throughput from 1 Gbps to 100 Mbps, for example, the occupied bandwidth can be reduced tenfold and a 10 times narrower receive filter can be employed. This will reduce the noise floor of the receiver by 10 dB and improve the receiver sensitivity by the same amount. This adaptive data rate technique allows links to provide a similar functionality to ACM while maintaining the advantages of lower order modulation.

7.4.4   Ring and Mesh Topologies

Section 6.3.1 showed several different topologies that can be used for building high capacity WiMAX and cellular networks (Figure 6.10). It was shown how ring and mesh configurations are able to reduce congestion and add resiliency over traditional hub-and-spoke network designs. One additional benefit of ring and mesh networks is that they can also improve the overall availability of the network.

Because hub-and-spoke networks have single points of failure, usually in the high data carrying aggregation links, critical paths often require 1+1 protected links, with parallel radios operating over parallel paths to provide redundancy in case of equipment failure. In a 1+1 configuration, both radios share the same transmission path, making both susceptible to the same rain fade outages. In the alternative ring or mesh architecture, there are at least two links connecting each node, each on a different path. Since the intense rain that causes link outages tends to be concentrated in well-defined 2–3-km diameter cells (see Section 7.3.4), the path diversity inherent in ring and mesh networks provides added robustness to the network. If the rain cell causes one link to fade in a ring or mesh network, no base station nodes will experience outages because traffic will be rerouted around the network, avoiding the problematic connection. Only if the rain cell is concentrated above the base station, causing all links connected to that node to fail, will the node become inoperable. For this reason, wireless links configured in ring or mesh networks can be installed with less weather availability per link than in hub-and-spoke configurations to maintain the same overall network availability. This means longer-distance links can be tolerated in ring and mesh topologies.

7.5 Summary

This chapter combines all the information from previous chapters and shows how high capacity 6–40-GHz microwave, 60-GHz millimeter-wave, and 70/80-GHz millimeter-wave radio links can be characterized and how wireless circuits can be designed.

A number of examples are given to show how high capacity microwave and millimeter-wave radios perform in real life. Using equipment parameters typical for commercially available equipment, it is shown that the high atmospheric attenuation and tight regulatory restrictions will always result in 60-GHz wireless being a several hundred meters or less short-distance technology. The 70/80 GHz offers similar data rate handling capabilities, but in a much friendlier atmospheric environment and with more flexible regulatory rules. For these reasons, typical 70/80 GHz radios can provide 1.5–2-km transmissions for 99.999% weather availability, and 5–7 km for 99.9% weather availability in most of the world. Microwave radios in the 6–40-GHz band provide the longest-distance high capacity wireless links. The 1.4-Gbps multichannel 23-GHz example considered offers about a 50% increase in distance over the 70/80-GHz radio. Such a link can achieve typically 2–3 km for carrier-class availability and about 10 km for lesser 99.9% weather availability. Knowledge of the link budget for the equipment and rain statistics for its intended location will allow tradeoffs of operating parameters such as output power, antenna size, and data throughput to achieve the important link engineering operational parameters of link distance and availability. A number of configuration and implementation techniques are available that can increase the robustness of a high data rate wireless link or network and allow link distances to be increased for a given performance criteria.

References

[1] Crane, R. K., "Prediction of Attenuation by Rain," *IEEE Trans. Commun.*, Vol. 28, No. 9, 1980, pp. 1717–1733.

[2] ITU-R P.837-4, "Characteristics of Precipitation for Propagation Modeling," 2003.

[3] Myers, W., "Comparison of Propagation Models," IEEE 802.16cc-99/13, 1999.

[4] WCA-PCG-7080-1, "Path Coordination Guide for the 71-76 and 81-86 GHz Millimeter Wave Bands," 2004.

List of Acronyms and Abbreviations

100GbE   100 Gbps Ethernet

10GbE   10 Gbps Ethernet

3G   third generation

3GPP   Third Generation Partnership Project

40GbE   40 Gbps Ethernet

4G   fourth generation

AC   alternating current

ACCP   adjacent channel copolarization

ACM   adaptive coding and modulation

ACMA   Australian Communications and Media Authority

ALC   automatic level control

AM   analog modulation

ARIB   Association of Radio Industries and Businesses

ASIC   application specific integrated circuit

ASK   amplitude shift keying

ATE   adaptive transversal equalizer

ATPC   automatic transmit power control

AV   audiovisual

BER   bit error rate

BiCMOS bipolar complimentary metal oxide semiconductor

BPSK   binary phase shift keying

BSC   base station controller

BTS   base transceiver station

BWRC   Berkeley Wireless Research Center

CCDP   cochannel dual-polarization

CCIR   Consultative Committee on International Radio

CDMA   code division multiple access

CE   Conformité   Européenne   (European   Conformity),   consumer electronics

CEPT   Conference of Postal and Telecommunications Administrations

CFR   Code of Federal Regulations

CMOS   complimentary metal oxide semiconductor

CORES   Commission Registration System

CPE   customer premise equipment

CPRI	common public radio interface
CW	continuous wave
DAS	distributed antenna system
dB	decibel
DC	direct current
DCDP	dual channel dual polarization
DFE	decision feedback equalizer
DoC	declaration of conformity
DQPSK	differential quadrature phase shift keying
ECC	Electronic Communications Committee
ECO	European Communications Office
E-field	electric field
EHF	extremely high frequency
EIRP	effective isotropic radiated power
EMC	electromagnetic compatibility
EN	European Norm
eNB	evolved node B
ESD	electrostatic discharge
ETSI	European Telecommunications Standards Institute
FCC	Federal Communications Commission
FDD	frequency division duplex

FEC	forward error correction
FFE	feed forward equalizer
FM	frequency modulation
FPGA	field programmable gate array
FRN	FCC registration number
FSK	frequency shift keying
FSO	free space optics
FTP	File Transfer Protocol
GaAs	gallium arsenide
GaN	gallium nitride
GbE	gigabit Ethernet
Gbps	gigabits per second
GEDC	Georgia Electronic Design Center
GHz	gigahertz
GPS	Global Positioning System
GSM	Groupe Spécial Mobile, global system for mobile communications
H	horizontal
HAPS	high altitude platform systems
HD	high definition
HDFS	high density fixed service
HDMI	high-definition multimedia interface

HDTV	high-definition television
H-field	magnetic field
HPA	high power amplifier
HPBW	half-power beamwidth
HRP	high-rate PHY
HSSG	Higher-Speed Study Group
HTML	Hypertext Markup Language
I	in-phase
IC	Industry Canada
IDU	indoor unit
IEC	International Electrotechnical Commission
IEEE	Institute of Electrical and Electronics Engineers
IETF	Internet Engineering Task Force
IF	intermediate frequency
InP	indium phosphide
IP	Internet Protocol
IQ	in-phase/quadrature
ISI	intersymbol interference
ISO	International Organization for Standardization
ITU	International Telecommunication Union
ITU-R	ITU Radiocommunication Sector

kbps	kilobits per second
LAG	link aggregation
LAN	local area network
LDPC	low density parity check
LIU	line interface unit
LNA	low noise amplifier
LO	local oscillator
LOS	line of sight
LRP	low-rate PHY
LTE	long-term evolution
MAC	medium access control
Mbps	megabits per second
MHz	megahertz
MIMO	multiple-input multiple-output
MMIC	microwave monolithic integrated circuit
MPA	medium power amplifier
MSC	mobile switching center
MTBF	mean time between failures
MTTR	mean time to repair
NEBS	network equipment building system
NLOS	nonline of sight

NMS	network management system
NTIA	National Telecommunications and Information Administration
OBSAI	Open Base Station Architecture Initiative
ODU	outdoor unit
OET	Office of Engineering and Technology
Ofcom	Office of Communications
OFDM	orthogonal frequency division multiplex
OOB	out of band
OOK	on-off keying
OQPSK	offset quadrature phase shift keying
OSI	Open System Interconnection
PAN	personal area networks
PDH	plesiochronous digital hierarchy
PHEMT	pseudomorphic high electron mobility transistor
PHY	physical layer
PLL	phase locked loop
PM	phase modulation
PSK	phase shift keying
PTMP	point to multipoint
PTP	point to point
PWE3	pseudowire emulation edge to edge

Q	quadrature
QAM	quadrature amplitude modulation
QoS	quality of service
QPSK	quadrature phase shift keying
R&TTE	Radio Equipment and Telecommunications Terminal Equipment
RAN	radio access network
RF	radio frequency
RFU	radio frequency unit
RGB	red green blue
RIC	radio interface capacity
RNC	radio network controller
RPE	radiation pattern envelope
RRH	remote radio head
RSL	received signal level
RTPC	remote transmit power control
SAN	storage area network
SDH	synchronous digital hierarchy
SDI	serial digital interface
SFP	short form pluggable
SHF	super high frequency
Si	silicon

SiGe	silicon germanium
SMPTE	Society of Motion Picture and Television Engineers
SONET	synchronous optical networking
SPU	signal processing unit
TCB	Telecommunications Certification Body
TCF	technical construction file
TDD	time division duplex
TDM	time division multiplex
TEM	transverse electromagnetic
THz	terahertz
TR	transmit receive
TRA	Telecommunications Regulatory Authority
UHF	ultrahigh frequency
UL	Underwriters Laboratory
ULS	Universal Licensing System
UTP	unshielded twisted pair
UWB	ultrawideband
V	vertical
VCO	voltage controlled oscillator
VHF	very high frequency
VoIP	Voice over IP

WGA	Wireless Gigabit Alliance
WiFi	wireless fidelity
WiGig	Wireless Gigabit Alliance
WiMAX	Worldwide Interoperability for Microwave Access
WLAN	wireless local area network
WPAN	wireless personal area network
WRC	World Radiocommunication Conference
WTB	Wireless Telecommunications Bureau
XPD	cross-polarization discrimination, cross-polarization distortion
XPIC	cross-polarization interference cancellation

About the Author

Jonathan Wells is the president of AJIS LLC (www.ajisconsulting.com), an independent consultancy specializing in the wireless technology and telecommunications industries. He has 20 years of experience in the commercial wireless market. Dr. Wells specializes in development and commercialization strategies for emerging microwave and millimeter-wave products and applications. He also provides expert witness and litigation support on a variety of wireless topics. Dr. Wells has been active on international regulatory committees, and has advised wireless agencies in Europe, the Middle East, Asia, and Australia. He is a regular conference speaker and is widely published and quoted in both academic and commercial literature.

Prior to founding AJIS, Dr. Wells held various roles with the GigaBeam Corporation where he was responsible for product management, technical marketing, and worldwide regulatory affairs. Before this, he was director of product development for Stratex Networks (now Aviat Networks) where he was responsible for the company's complete RF and microwave product development. Previously Dr. Wells ran the Wideband Product Division for Adaptive Broadband, an early adopter of unlicensed wireless access products.

Dr. Wells holds a B.Sc. in physics, a Ph.D. in millimeter-wave electronics, and an M.B.A. with a specialization in strategic R&D management. He is a senior member of the IEEE. Dr. Wells can be reached at jonathan@ajisconsulting. com.

Index

For further information on these and other Artech House titles, including previously considered out-of-print books now available through our In-Print-Forever® (IPF®) program, contact:

Artech House
685 Canton Street
Norwood, MA 02062
Phone: 781-769-9750
Fax: 781-769-6334
e-mail: artech@artechhouse.com

Artech House
16 Sussex Street
London SW1V 4RW UK
Phone: +44 (0)20 7596-8750
Fax: +44 (0)20 7630-0166
e-mail: artech-uk@artechhouse.com

Find us on the World Wide Web at: www.artechhouse.com